*Risk-sensitive
Optimal Control*

WILEY-INTERSCIENCE SERIES IN SYSTEMS AND OPTIMIZATION

Advisory Editors

Peter Whittle

Statistical Laboratory, University of Cambridge, 16 Mill Lane, Cambridge CB2 1SB, UK

Sheldon Ross

Department of Industrial Engineering and Operations Research, University of California, Berkeley, CA 94720, USA

GITTINS—Multi-armed Bandit Allocation Indices

WHITTLE—Risk-sensitive Optimal Control

Risk-sensitive Optimal Control

Peter Whittle

Statistical Laboratory, University of Cambridge

JOHN WILEY & SONS
Chichester · New York · Brisbane · Toronto · Singapore

Copyright © 1990 by John Wiley & Sons Ltd.
Baffins Lane, Chichester
West Sussex PO19 1UD, England

All rights reserved.

No part of this book may be reproduced by any means,
or transmitted, or translated into a machine language
without the written permission of the publisher.

Other Wiley Editorial Offices

John Wiley & Sons, Inc., 605 Third Avenue,
New York, NY 10158-0012, USA

Jacaranda Wiley Ltd, G.P.O. Box 859, Brisbane,
Queensland 4001, Australia

John Wiley & Sons (Canada) Ltd, 22 Worcester Road,
Rexdale, Ontario M9W 1L1, Canada

John Wiley & Sons (SEA) Pte Ltd, 37 Jalan Pemimpin 05-04,
Block B, Union Industrial Building, Singapore 2057

Library of Congress Cataloging-in-Publication Data

Whittle, Peter.
    Risk-sensitive optimal control / Peter Whittle.
      p. cm.—(Wiley-Interscience series in systems and
optimization)
    Includes bibliographical references.
    ISBN 0 471 92622 1
    1. Control theory. 2. Mathematical optimization. I. Title.
II. Series.
QA402.3.W49   1990   89-28165
629.8'312—dc20   CIP

British Library Cataloguing in Publication Data

Whittle, Peter
    Risk-sensitive optimal control.
    1. Optimisation. Applications of variational methods
    I. Title
    515'.64
    ISBN 0 471 92622 1

Typeset by Asco Trade Typesetting Ltd., Hong Kong.
Printed in Great Britain by Courier International, Tiptree, Essex

We know there is both good and wicked chance:

But this false world, with many a double cast,

In it there's nought but endless variance.

> Blind Harry, the Minstrel
> (*The Story of Wallace Wight*)

It is better to debate a question without settling
it, than to settle a question without debating it.

Joseph Joubert, *Pensées*
(The Oxford Dictionary of Quotations, 1979)

Contents

Preface ix

Chapter 1 Models and intentions 1

Part I LQG theory summarized 13

Chapter 2 LQG structure: certainty equivalence 15
Chapter 3 The Markov case: control rules 31
Chapter 4 The Markov case: state estimation 59
Chapter 5 The complete dualization of past and future 67

Part II Risk sensitivity: the LEQG formulation 77

Chapter 6 The risk-sensitive certainty-equivalence principle 79
Chapter 7 The Markov case: future stress and control 93
Chapter 8 The Markov case: past stress and estimation 107
Chapter 9 The infinite horizon: limits, breakdown points and policy improvement 113

Part III The path-integral (Hamiltonian) approach 127

Chapter 10 Path-integral methods: the formalism 129
Chapter 11 The Markov case: recursions and factorizations 147
Chapter 12 Higher-order models: the general path-integral formalism 161
Chapter 13 Canonical factorization in the control context 173
Chapter 14 The recoupling of past and future 185
Chapter 15 Continuous time: the path-integral formalism 189
Chapter 16 Continuous time: control optimization 199

Part IV Connections and variations   205

Chapter 17 The relationship of the LEQG criterion to H_∞ and
            entropy criteria   207
Chapter 18 Variants   221

Appendix 1 Abbreviations   231
Appendix 2 Notation and a list of symbols   233
Appendix 3 Optimal estimation and the Gauss–Markov theorem   237
Appendix 4 The Hamiltonian formalism   239

References   241

Index   245

Preface

This work has two major themes. One is that of risk-sensitive control, in that the quadratic cost function of the standard LQG (linear/quadratic/Gaussian) treatment is replaced by the exponential of a quadratic, giving the so-called LEQG formulation. The effect of the generalization is to provide an extra tempering parameter, which can be said to set the level of the optimizer's degree of optimism or pessimism—i.e. his degree of belief that random events will work out to his advantage or disadvantage.

Formally viewed, the extension is very interesting. All the familiar LQG theory has an LEQG analogue, although a reorientation of ideas is sometimes needed before one can see it. Once one has achieved this reorientation, then the LQG ('risk-neutral') theory appears, not only as a special case but almost as a degenerate one. For example, the risk-sensitive certainty-equivalence principle yields exactly the stochastic maximum principle, couched wholly in terms of observable and computable quantities, which has been sought for years. Also, as has been realized by Glover and Doyle, the H_∞ and minimum entropy criteria, which have awoken so much interest in recent years, amount exactly to an infinite horizon version (more exactly, an average-optimal version) of LEQG theory, despite the great difference in starting point.

The second theme is the so-called path-integral or Hamiltonian formulation. The use of recursive methods of optimization reduces the optimal control problem to one of solving a Riccati equation. However, it has been known for years that there is an elegant formulation of optimal LQ control wholly in terms of linear operators, a canonical operator factorization replacing the solution of a Riccati equation. This approach generalizes naturally to higher-order (non-Markov) models. However, it also covers the LEQG formulation in the most natural way. Appeal to the risk-sensitive certainty-equivalence principle leads to a restatement of the control-optimization problem as the extremization of a quadratic path integral. The linear stationarity equations have an appealing Hamiltonian form, and the only entry in the matrix operator of this system which is vacant in the LQ case is neatly filled by the risk-sensitivity parameter.

Let me ease the task of reviewers (since reviewers are also authors and authors also reviewers) by summarizing what I believe to be the positive and the deficient aspects of this book. For the positive features, there are those listed above: the risk-sensitive certainty-equivalence and separation principles (Chapter 6), the consequent extension of the conventional LQG treatment (Chapters 7 and 8) and the path-integral formulation (Chapters 10–16). There are others: the exploitation of the equivalence between policy improvement and the Newton–Raphson method to yield fast guaranteed iterative methods of canonical factorization (Sections 3.8, 9.4, 11.5, 12.4, 12.5, 13.4 and 15.4), the deduction of the natural relationship between the value function and the canonical factorization (Sections 12.3 and 15.3), the deduction of the form of the canonical factors (Sections 11.3, 13.1 and 16.3) and the integration of LEQG theory with H_∞ and minimum-entropy methods (Chapter 17). Finally, the whole path-integral formalism, with its associated Hamiltonian structure and stochastic maximum principle, has a natural extension to the non-LQG case if an exponential-of-cost criterion is adopted and if large-deviation theory is applicable (see Section 18.5, added in proof).

On the debit side, I am conscious of the fact that there should be much more in the way of numerical examples, both to carry conviction that the methods suggested really offer something and to explore the range of their validity as the initial trial solution and the value of the risk-sensitivity parameter are varied. Also, there is theory yet to complete. The sufficient conditions given for the existence of canonical factorizations (appealing to conditions of the type of controllability and deviation-sensitivity, see Sections 9.1, 9.2 and 13.3) are far from the best possible, and do not have the natural form for higher-order risk-sensitive models which must surely exist. Also the problem of matching past and future optimizations in the higher-order case (Chapter 14) can surely be reduced further.

In mitigation, I can only enter the lesser plea that one reaches one's current personal limit at a certain point, and the greater plea: that the linear model again demonstrates its will-o'-the-wisp quality. That is, this simple model seems to have infinite depth, and yields only to reveal further mysteries. Its known theory becomes more extensive and definite with time, but somehow never definitive. However, I am convinced that the path followed in this text is the natural and rewarding one, and I should like nothing so well as to see others continue it.

For all but the last six months of work on this book I was supported by the Esso Petroleum Company, through tenure of the Churchill Chair. For the final six months I was supported as a Senior Fellow by the Science and Engineering Research Council. It is a pleasure to acknowledge my indebtedness to these bodies.

P. WHITTLE
Cambridge

CHAPTER 1

Models and Intentions

For these first few sections we take some matters of convention and notation for granted; see Section 1.5 for directions.

1.1 THE CLASSIC MODEL

The elements needed to characterize a control optimization problem are (1) a specification of the dynamics of the controlled process, (2) a specification of which quantities are observable at a given time and (3) an optimization criterion.

We shall use x, u and y to denote process variable, control variable and observation, respectively, the values of these variables at a moment t in discrete time being denoted x_t, u_t and y_t. (By 'process variable' we mean whatever variables are needed for a quantitative dynamic model of the process; see Exercise 1.1.1 for a more exact characterization.)

In our context we shall assume throughout that x, u and y are finite-dimensional vector variables, of dimensions n, m and r, respectively. One of the simplest non-trivial models for controlled stochastic plant dynamics prescribes x as obeying the *plant equation*

$$x_t = Ax_{t-1} + Bu_{t-1} + \varepsilon_t \tag{1.1}$$

Here A and B are fixed matrices of appropriate size and ε is vector *white noise* (see Exercise 1.1.1(2). This model is special in that it is linear, Markov and time homogeneous. It is linear in that (1.1) is a linear difference equation, linearly driven by control u and noise ε. It is Markov in that the distribution of x_t conditional on process history $\{x_\tau, u_\tau; \tau < t\}$ is in fact conditioned only by the most recent values: (x_{t-1}, u_{t-1}). (This makes the process variable x in fact a *state variable* as far as the dynamics of the process are concerned.) By *time homogeneity* one means that the stochastic dynamics of the process do not vary with time; relation (1.1) is then time homogeneous in that A, B and the noise statistics are independent of time (see Exercise 1.1.4).

The process variable and process history are not, in general, observable. The quantities presumed known at time t are present and past observations $(y_t, y_{t-1}, \ldots)$ and past actions $(u_{t-1}, u_{t-2}, \ldots)$. A common assumption companion to equation (1.1) is that the observations y_t are generated by

$$y_t = Cx_{t-1} + \eta_t \tag{1.2}$$

where C is also a fixed matrix and (ε, η) jointly constitute vector white noise. That is, y_t is a noise-corrupted version of certain linear functions of x_{t-1}. There are formal reasons which make it natural to assume that y_t is an imperfect observation of x_{t-1} rather than of x_t. The random inputs ε and η are referred to as *process noise* and *observation noise*, respectively.

A *policy* π is a specification of the control u_t at time t as a function of the observables W_t at time t, for all relevant t. In the above case we would take W_t as being $(y_t, y_{t-1}, \ldots; u_{t-1}, u_{t-2}, \ldots)$.

Whereas the plant equation and the nature of the observables must be regarded as given, the optimizer can choose the control policy as he sees best. The choice of a control policy π completes the stochastic specification of the controlled process, and induces an expectation operator E_π.

The optimizer will, in general, choose π to minimize a criterion of the form $E_\pi(\mathbb{C})$. Here $\mathbb{C}$ is a *cost function*: a scalar function of the process path $\{x_t, u_t\}$ which measures the 'cost' or 'penalty' incurred because of the deviation of this path from some desired course. For a process with state structure which is to be optimized over a finite time interval $0 \le t \le h$ a common form of cost function is

$$\mathbb{C} = \sum_{t=0}^{h-1} c(x_t, u_t, t) + \mathbb{C}_h(x_h) \tag{1.3}$$

where the *instantaneous cost* $c(x, u, t)$ and the *terminal cost* $\mathbb{C}_h$ are chosen to provide an appropriate penalty measure. There is a very special theory for the case when this cost function is quadratic in the x and u arguments. In particular, if one is trying to penalize deviations from $x = 0$, $u = 0$ (i.e. to *regulate* to the set point $(0, 0)$) then one might choose

$$c(x, u, t) = x'Rx + u'Sx + x'S'u + u'Qu = \begin{bmatrix} x \\ u \end{bmatrix}' \begin{bmatrix} R & S' \\ S & Q \end{bmatrix} \begin{bmatrix} x \\ u \end{bmatrix} \tag{1.4}$$

$$\mathbb{C}_h(x) = x'\Pi_h x \tag{1.5}$$

with appropriately chosen matrices R, S, Q and Π_h.

For optimization over a finite time interval the terminal cost $\mathbb{C}_h(x_h)$ in a sense supplies a *terminal condition*, in that it specifies the cost of terminating with a given value of terminal state x_h. One correspondingly needs an *initial condition*: a specification of the information W_0 available at time 0. For a process with state structure this will amount to the specification of the distribution of x_0; the form of this *prior distribution* constitutes W_0.

A common assumption is that this distribution is Gaussian with prescribed mean vector and covariance matrix, $\hat{x}_0$ and V_0, say. That is, $\hat{x}_0$ is regarded as the best estimate $E(x_0 | W_0)$ of x_0 that can be formed on initial information, and the *estimation error* $\Delta_0 = \hat{x}_0 - x_0$ is assumed normal with zero mean and covariance matrix V_0.

If one adds to assumptions (1.1)–(1.5) the hypothesis that all the noise variables

Δ_0 and (ε_t, η_t) $(0 < t \leq h)$ are jointly normally distributed, then one has the classic LQG model, a state-structured time-homogeneous form of the general LQG model to be formulated in Chapter 2. The term LQG is a convenient abbreviation of linear/quadratic/Gaussian. The linear aspect is that the plant equation and observation relation are linear (in x and u); the quadratic aspect is that the cost function $\mathbb{C}$ is quadratic in these variables, and the Gaussian aspect is that all injected noise variables (initial estimation errors included) are jointly normal.

LQG models have a particularly rewarding structure and a very complete theory. As we have already emphasized, the model specified by assumptions (1.1)–(1.5) is very special as an LQG model. In particular, state structure is something quite separate from LQG structure. The principal simplification induced by LQG structure is that the optimal control values turn out to be linear functions of the observations. As far as the determination of these linear rules are concerned, an important secondary consideration is the validity of a *certainty-equivalence principle*, to be explained in Chapter 2.

One principal aim of this work is to show that the LQG model can be embedded in a larger class of models: the LEQG models of Section 1.2. This provides a generalization so powerful that many aspects of the simple LQG formulation can only be truly appreciated in the LEQG setting. Furthermore, the treatment of this larger class leads to a natural integration of the various optimization techniques.

Everything we do in discrete time has a natural analogue in continuous time, which is usually sketched without too much delay. The continuous analogue of the model (1.1)–(1.5) is indicated in Exercise 1.1.3.

1.1.1 Exercises and comments

(1) It is the essence of a process variable that (i) it should be *complete*, in that it describes all aspects of the system which are of interest, and (ii) *autonomous*, in that, in the discrete-time deterministic case, it obeys a recursion of the form

$$x_t = a(X_{t-1}, U_{t-1}, t)$$

Here X_t is the x-history up to time t, $\{x_\tau; \tau \leq t\}$, etc., so one is saying that x_t should be determined by process and control histories before time t. In the stochastic case, one will demand that the conditional probabilities $P(x_t | X_{t-1}, U_{t-1})$ be well defined. (See Whittle (1982) p. 150 for a statement of the conditions which the variables of a problem must satisfy if the temporal optimization problem is to have the character which one intuitively expects.)

(2) We shall usually assume that the noise sequences have zero mean. If we suppose that $\text{cov}(\varepsilon_t) = N$ then a second-order characterization of the white-noise property is that $E(\sum_t \alpha_t' \varepsilon_t) = 0$, $E(\sum_t \alpha_t' \varepsilon_t)^2 = \sum_t \alpha_t' N \alpha_t$ for any set of n-vectors α_t for which this last sum is convergent. If one makes the stronger assumption (which we do for most of the text) that $\{\varepsilon_t\}$ is Gaussian (i.e. that the ε_t are jointly

normally distributed) then one can assert that

$$E\exp\left(\sum_t \alpha_t' \varepsilon_t\right) = \exp\left(\frac{1}{2}\sum_t \alpha_t' N \alpha_t\right) \quad (1.6)$$

(3) In continuous time we shall usually write $x(t)$ rather than x_t, etc. The analogue of the plant equation (1.1) is

$$\dot{x} = Ax + Bu + \varepsilon \quad (1.7)$$

where $\dot{x} = dx/dt$ and ε is continuous-time white noise. The statistics of ε are described by the continuous analogue of equation (1.6):

$$E\exp\left(\int \alpha(t)' \varepsilon(t)\,dt\right) = \exp\left(\frac{1}{2}\int \alpha(t)' N \alpha(t)\,dt\right)$$

which implies that ε itself has infinite covariance. The legitimizations of this point are well known, e.g. that ε is understood as an ideal limit of finite-variance processes in a way which defines it self-consistently as a *generalized process* (see e.g. Gihman and Skorokhod, 1971).

The observation relation (1.2) takes the non-differential form

$$y = Cx + \eta \quad (1.8)$$

The cost function (1.3) modifies to the obvious integral analogue

$$\mathbb{C} = \int_0^h c(x, u, t)\,dt + \mathbb{C}_h(x(h)) \quad (1.9)$$

with instantaneous cost rate c and terminal cost $\mathbb{C}_h$ still having the definitions (1.4) and (1.5).

(4) One should distinguish between *time homogeneity* and *time invariance*. An optimization problem is time homogeneous if none of the plant equation, the observation relation or the instantaneous cost are explicitly time dependent. One is in the time-invariant case, when the whole optimization problem is invariant to an arbitrary time translation. This will require time homogeneity, termination rules and costs which are time independent, and a history of operation into the remote past. If, as one would wish, an optimal control rule should also exist which is *stationary* (i.e. not explicitly time dependent) then additional regularity hypotheses will, in general, be needed (e.g. that stationary stabilizing controls exist).

1.2 RISK SENSITIVITY AND THE LEQG FORMULATION: OPTIMISM AND PESSIMISM

Suppose that, instead of choosing policy π to minimize $E_\pi(\mathbb{C})$, one chooses it to minimize

$$\gamma_\pi(\theta) = -(2/\theta)\log(E_\pi \exp(-\theta \mathbb{C}/2) \quad (1.10)$$

1.2 RISK SENSITIVITY AND THE LEQG FORMULATION

where θ is a parameter which we shall term the *risk-sensitivity parameter*. If the assumptions are otherwise LQG, then we shall term this an *LEQG formulation*, the EQ now being accepted as standing for 'exponential of quadratic'.

Some notion of the nature of this modified criterion is obtained by expanding $\gamma_\pi(\theta)$ in powers of θ for θ small. One obtains

$$\gamma_\pi(\theta) = E_\pi(\mathbb{C}) + (\theta/4)\operatorname{var}_\pi(\mathbb{C}) + O(\theta^2) \tag{1.11}$$

where $\operatorname{var}_\pi(\mathbb{C})$ is the variance of $\mathbb{C}$ under policy π. That is, for $\theta = 0$ we have the simple *risk-neutral* criterion $E_\pi(\mathbb{C})$, whereas variability of the random cost $\mathbb{C}$ about this expected value is seen as advantageous if $\theta > 0$, disadvantageous if $\theta < 0$.

More specifically, if $\theta > 0$ then one seeks to maximize the expectation of $\exp(-\theta\mathbb{C}/2)$, a convex decreasing function of $\mathbb{C}$. This criterion throws relatively more weight on the smaller values of $\mathbb{C}$ as θ increases, less on the larger values. That is, while optimizers in general wish to reduce cost, they are (relative to the risk-neutral case) more concerned by the frequent occurrence of moderate values of $\mathbb{C}$ than by the occasional large value. Economists would term such optimizers *risk seeking*; we may well call them *optimistic* in their disregard of worst cases relative to typical or best cases.

For $\theta < 0$ the situation is indeed reversed; optimizers then seek to minimize the expectation of $\exp(-\theta\mathbb{C}/2)$, a convex increasing function of $\mathbb{C}$. This criterion shifts concern to the larger values of $\mathbb{C}$ as θ decreases (and so increases numerically). Optimizers then becomes increasingly concerned to control worst cases rather than typical or best cases. Economists would term such optimizers *risk-averse*; we may well call them *pessimistic*.

These characterizations would have held if we had considered other criteria of the form $E_\pi(L(\mathbb{C}))$, where the function L rescales the cost axis in some way. (Here L is the negative of what economists would term a *utility function*, so perhaps they would term it a *disutility function*: see Exercise 1.2.1(1). However, the point about the particular choice (1.10), with L exponential, is that the exponential of a quadratic form is the very pattern of a Gaussian density function. The theory based on this choice is then found to be both explicit and interpretable.

The parameter θ provides a valuable extra degree of freedom in formulation, which allows one to temper the optimization by an awareness of variability; to tune the optimizer to some point on the scale from optimism to pessimism. Indeed, a modest degree of optimism might be desirable, in that the conventional LQG treatment is often held to give excessive weight to the occasional large value of $\mathbb{C}$.

The exponential disutility function has another property: to a first approximation it decouples the effects of the mean value of $\mathbb{C}$ and the variability of $\mathbb{C}$. See Exercise 1.2.1(3).

1.2.1 Exercises and comments

(1) If $\mathbb{C}$ is the cost, but the optimizer chooses to minimize $E_\pi(L(\mathbb{C}))$, then $L(\mathbb{C})$ must measure the perceived cost in some sense; a *disutility*. One will usually expect the function L to be monotone non-decreasing. If it is convex, then, by Jensen's inequality,

$$EL(\mathbb{C}) \geq L(E\mathbb{C}) \qquad (1.12)$$

with equality if $\mathbb{C}$ shows no variability, i.e. simply takes the fixed value $E(\mathbb{C})$. A convex disutility function thus penalizes variability, and indicates a risk-averse or pessimistic attitude on the part of the optimizer. If the function L is concave then inequality (1.12) is reversed, and variability is seen as advantageous. This indicates a risk-seeking or optimistic attitude on the part of the optimizer.

(2) Let L^{-1} be the function inverse to L. In basing the criterion upon a disutility $L(\mathbb{C})$ it is advantageous to take the actual criterion as $L^{-1}(E_\pi(L(\mathbb{C})))$, so that one has transformed back into the original cost scale. A non-singular linear transformation is then seen as irrelevant, in that $L^{-1}(E_\pi(L(\mathbb{C}))) = E_\pi(\mathbb{C})$. This corresponds to the fact that L and E_π then commute.

Note that in the LEQG case $L(\mathbb{C}) = \exp(-\theta\mathbb{C}/2)$. One then minimizes the criterion $L^{-1}(E_\pi(L(\mathbb{C})))$, whether θ is positive, negative or zero.

(3) Set $E_\pi(\mathbb{C}) = m$, $\text{var}_\pi(\mathbb{C}) = v$. Show that the general version of relation (1.11) is

$$\gamma = L^{-1} E_\pi L(\mathbb{C}) = m + \frac{L''(m)}{L'(m)} \frac{v}{2} + \cdots$$

where the terms neglected are of order $E(\mathbb{C} - m)^j$ for $j > 2$. If one demands that L be such that the sensitivity of the criterion γ to variability of $\mathbb{C}$ be the same for all levels of $m = E(\mathbb{C})$, then, to first order, one demands that $L''(m)/L'(m)$ be constant in m. Show that this implies that $L(\mathbb{C})$ is, to within a linear transformation, an exponential function of $\mathbb{C}$. See Pratt (1964), Arrow (1971)

1.3 HIGHER-ORDER MODELS

The plant equation (1.1) could be regarded as a first-order stochastic difference equation, linear and with constant coefficients. It is then natural to consider higher-order versions:

$$\sum_{j=0}^{p} A_j x_{t-j} + \sum_{j=1}^{q} B_j u_{t-j} = \varepsilon_t \qquad (1.13)$$

say, where ε is still vector white noise.

Economists certainly find themselves compelled to consider such 'distributed lag' models. This is probably because, in economic contexts, not only could one not hope to observe the state variable, but one would also feel unsure in asserting that any given set of variables would constitute an adequate state description. It is then better to allow that the process variables adopted could follow some higher-order scheme.

If p and q are finite then it is indeed possible to choose new variables in terms of which model (1.13) is brought to the Markov form (1.1); see Exercise 1.3.2. However, such a transformation in fact loses some of the special structure of the process, in that the Markov representation achieved will be special as a Markov process; see Exercise 1.3.2 again. One will not exploit this special structure if one always normalizes to the Markov form.

In fact, there are methods available for the treatment of models such as (1.13) (generating function and Fourier/Laplace methods) which do not rely at all upon Markov structure, but rather upon the fact that the dynamics expressed by (1.13) are linear and time homogeneous. These methods yield a particularly clean and general treatment in the stationary case. That is, when the time interval over which one optimizes extends so far in both directions that one attains equilibrium conditions, initial information and terminal costs being too remote to affect estimation or control optimization, respectively.

One can say that there are two distinct approaches to the optimization of LQG models which have proved fruitful. One is the assumption of stationarity, with a consequent exploitation of Fourier/Laplace methods. The other is the assumption of state structure, with a consequent exploitation of recursive relationships. There is no opposition between the two approaches, neither of which includes or totally excludes the other. One must simply choose that which is appropriate in a given case. The exploitation of stationarity followed on somewhat from Wiener's classic work on optimal prediction (Wiener, 1949) and was pursued from quite an early date by a number of authors: e.g. Newton (1952), Newton *et al.* (1957), Holt *et al.* (1960) and Whittle (1963). (It should also be emphasized, however, that the methods we shall use are distinct from Wiener's in some significant respects; see the comments in Section 1.4 and Exercise 5.3.1).

The state-variable approach was particularly pioneered in this context by Kalman (Kalman, 1960, 1963; Kalman and Bucy, 1961) and has tended to carry all before it ever since. This is understandable; a state formulation is felt to correspond to a physically complete formulation, and recursive methods make for natural algorithms. However, if one is willing to postulate time invariance then the Fourier methods have a power and elegance which in fact surpasses even that usually attributed to them. It is hoped that the reader will be convinced of this in Part III.

It is useful to introduce the backward translation operator $\mathcal{T}$, having the effect

$$\mathcal{T} x_t = x_{t-1} \qquad (1.14)$$

and so the iterated effect

$$\mathcal{T}^j x_t = x_{t-j} \qquad (j = 0, 1, 2, \ldots) \qquad (1.15)$$

We can then write the plant equation (1.13) as

$$\mathscr{A}x + \mathscr{B}u = \varepsilon \tag{1.16}$$

where the time argument is understood, and $\mathscr{A}$ and $\mathscr{B}$ are time-invariant linear operators which can be written

$$\mathscr{A} = \sum_j A_j \mathscr{T}^j$$
$$\mathscr{B} = \sum_j B_j \mathscr{T}^j \tag{1.17}$$

Correspondingly, we can consider a generalized form of the observation relation (1.2):

$$y + \mathscr{C}x = \eta \tag{1.18}$$

where

$$\mathscr{C} = \sum_j C_j \mathscr{T}^j \tag{1.19}$$

The particular pair (1.1) and (1.2) then correspond to the choices $\mathscr{A} = I - A\mathscr{T}$, $\mathscr{B} = -B\mathscr{T}$ and $\mathscr{C} = -C\mathscr{T}$.

This formalism also extends to the case of continuous time with the differential operator

$$\mathscr{D} = d/dt \tag{1.20}$$

replacing $\mathscr{T}$. The continuous-time Markov model (1.7) and (1.8) can then be written in the general form (1.16) and (1.18) if we take $\mathscr{A} = \mathscr{D} - A$, $\mathscr{B} = -B$ and $\mathscr{C} = -C$. More generally, we would assume (1.16) and (1.18) with the higher-order structure

$$\mathscr{A} = \sum_j A_j \mathscr{D}^j, \qquad \mathscr{B} = \sum_j B_j \mathscr{D}^j, \qquad \mathscr{C} = \sum_j C_j \mathscr{D}^j \tag{1.21}$$

For economy of notation we denote the coefficient matrices by A_j, etc. here as well as in the discrete case, (1.17) and (1.19). However, there is no very immediate relation between the two sets of coefficients. For example, the assertion that A_0 should be non-singular in the discrete-time case (see Exercise 1.3.1) has as continuous-time analogue the assertion that the matrix of coefficients of highest-order derivatives should be non-zero. That is, if A_i has jkth element $a_{jk}^{(i)}$ and derivatives of the jth element of x occur up to order r_j in the plant equation, then the matrix with jkth element $a_{jk}^{(r_k)}$ should be non-singular.

1.3.1 Exercises and comments

(1) If relation (1.13) is to constitute a forward recursion for x_t (i.e. to determine x in terms of past x and u and present and past ε) then we require that A_0 be non-singular. One then usually adopts the normalization $A_0 = I$.

If we regard u_t as being chosen *after* x_t is determined, then causality will require that $B_0 = 0$.

(2) *Reduction to state form.* Consider the plant equation

$$\sum_{j=0}^{p} A_j x_{t-j} = B u_{t-1} + \varepsilon_t$$

which is of order p in its dependence upon the x-past, but first order in its dependence upon the u-past. If we define an augmented variable

$$\tilde{x}_t = \begin{bmatrix} x_t \\ x_{t-1} \\ \vdots \\ x_{t-p+1} \end{bmatrix}$$

then this is a state variable, in that it satisfies the first-order plant equation

$$\tilde{x}_t = \tilde{A}\tilde{x}_{t-1} + \tilde{B} u_{t-1} + \tilde{\varepsilon}_t \qquad (1.22)$$

with

$$\tilde{A} = \begin{bmatrix} -A_1 & -A_2 & \cdots & -A_{p-1} & -A_p \\ I & 0 & \cdots & 0 & 0 \\ 0 & I & \cdots & 0 & 0 \\ \cdots & \cdots & \cdots & \cdots & \cdots \\ 0 & 0 & \cdots & I & 0 \end{bmatrix} \qquad (1.23)$$

$$\tilde{B} = \begin{bmatrix} B \\ 0 \\ 0 \\ \vdots \\ 0 \end{bmatrix} \qquad \tilde{\varepsilon}_t = \begin{bmatrix} \varepsilon_t \\ 0 \\ 0 \\ \vdots \\ 0 \end{bmatrix}$$

The matrix $\tilde{A}$ is often termed the *companion matrix* of the operator $\mathscr{A} = \sum A_j \mathscr{T}^j$. While it is true that the special form (1.23) of $\tilde{A}$ is lost under a linear transformation of $\tilde{x}$, it also true that the model (1.22) is special, in that both control and noise are injected into this np-dimensional process in some n-dimensional way.

The more general model (1.13) can be reduced to state form by taking x_{t-j} ($0 \le j < p$) and u_{t-j} ($0 < j < q$) as the components of a state variable, and analogous assertions hold.

(3) Note the formal relation $\mathscr{T} = \exp(-\mathscr{D})$.

1.4 INTENTIONS

The point of the text is the following. Ultimately, we consider the more general risk-sensitive or LEQG formulation of Section 1.2. The 'risk-sensitive certainty-equivalence principle' of Chapter 6 then reduces the deduction of control and estimation rules implicit in policy optimization to the seeking of the stationary point of a certain quadratic form, the 'path integral'. In the conventional Markov (i.e. state-structured) case this reduction then yields a treatment of control optimization generalizing the usual LQG treatment; see Chapters 7–9. However, more generally, the stationarity conditions for the path integral yield a set of linear equations with associated operator $\Phi(\mathcal{T})$ of a particularly natural and symmetric form. Canonical factorization of Φ yields the optimal control relations in closed-loop form (i.e. in a form which is optimal, whatever past control policy may have been). The factorization gives the control rule immediately in the infinite-horizon case, somewhat less immediately in the finite-horizon case. If the process variable is not perfectly observed then there are two operators to be factorized, corresponding roughly to deduction of estimates and of control rule.

It is in this sense that that path-integral methods and the risk-sensitive formulation fit neatly together. It should be noted that the path-integral approach is valid without assumptions of either state structure or time homogeneity. However, we assume time homogeneity so as make most use of the Fourier methods which then become available.

The approach constitutes a generalization of the maximum principle which in fact reveals points that the conventional risk-neutral case does not. In this sense, the risk-neutral case indeed appears degenerate. For instance, the co-state variables (Lagrangian multipliers) of the maximum principle are found to have a natural interpretation in the risk-sensitive case: they are related to the estimates of process and observation noise. One may indeed say that the risk-sensitive certainty-equivalence principle constitutes the proper statement of a 'stochastic maximum principle' (in the LQG case) which has so long been sought. Although the model is stochastic, the principle leads to equations which have solutions wholly in terms of current observables. See also Section 18.5.

The canonical factorizations of which we spoke are not those associated with calculation of the Wiener filter. If we consider the estimation/prediction aspects of the problem then the Wiener filter is a least-square (LS) one, leading to equation systems with covariance matrices as coefficients. The path-integral formulation to which we are led corresponds rather to a maximum likelihood (ML) estimation (at least, it reduces to this in the risk-neutral case), and the consequent equations systems have information matrices as coefficients. It follows from the Gauss–Markov theorem (Section 2.4) that the LS and ML approaches lead to identical estimates and predictors in the risk-neutral case. However, even in this case, the ML approach has substantial advantages: e.g. one has to factorize only a polynomial rather than a rational function, and the recursions which hold between predictors of different leads or based upon different observables are immediately apparent (Section 5.3.). In the more general risk-sensitive case the only natural route is by a version of the ML approach.

1.4 INTENTIONS

The calculation of a canonical factorization then becomes the final computational/analytic step, equivalent to the solution of a Riccati equation in recursive treatments of the state-structured case. There is an enormous literature on canonical factorization, but a particularly appealing group of methods, showing second-order convergence, are those based on Newton–Raphson methods (see Section 12.6 for references). These methods are found to provide the natural computational tool in our context, because we observe (see Sections 3.8 and 12.4) that, in the LQG and LEQG contexts, the Newton–Raphson method is exactly equivalent to the method of policy improvement. The method thus has a natural variational basis in this context.

There is yet one further fascinating turn to the story. A control criterion which has aroused great interest in recent years is that of optimization in the H_∞ norm. Suppose we write the plant equation in the form

$$\mathscr{A}x + \mathscr{B}u = \varepsilon = Wv$$

where v is regarded as a disturbing input. It is not regarded as necessarily stochastic, but the relation with our stochastic formulation will be found to be that $N = \mathrm{cov}(\varepsilon)$ is to be identified with WW'. So, if v were regarded as stochastic, it would be standard white noise, with $\mathrm{cov}(v) = I$.

Suppose there is a vector χ, a linear function of x and u, which is essentially regarded as 'system error' in that the aim of optimal control is to make χ small in some sense. Suppose stabilizing policies exist for which system error is related to disturbing input by $\chi_t = \sum_{j=0}^{\infty} G_j v_{t-j}$. Then $G(z) = \sum_j G_j z^j$ is the discrete-time transfer function of the system from input v to output χ. This will be analytic at least in $|z| \leq 1$ if the policy is indeed stabilizing.

The H_∞ criterion is then that the control rule should be chosen so as to make the system as insensitive to disturbing inputs as possible, in that the maximal frequency response of χ to v should be minimized. This is formalized as requiring that the maximal singular value of the response function $G(z)$ on $|z| = 1$ should be minimal. There is now an extensive theory of H_∞ control, and H_∞-optimal controls can be determined.

Glover and Doyle (1988) have demonstrated that the LEQG and H_∞ theories are intimately related, despite the fact that the criteria seem so radically different. Moreover, both the motivation and the analysis of the LEQG case throw great light upon the H_∞ case. We give a full discussion of these matters in Chapter 17.

1.4.1 Exercise and comments

(1) If G is the transfer function from v to χ then, in discrete time, we mean by this that $\chi = G(\mathscr{T})v$. For the particular case of exponential sequences we have $G(\mathscr{T})z^{-t} = G(z)z^{-t}$, and it is the function $G(z)$ of a scalar argument z that we think of as the transfer function.

In continuous time one has rather $\chi = G(\mathscr{D})v$, so that $G(\mathscr{D})e^{i\omega t} = G(i\omega)e^{i\omega t}$.

1.5 NOTATION AND STYLE

Conventions on notation are listed in Appendix 2. At this point we note merely the conventions already introduced without comment: that matrices are indicated by italic capital letters (e.g. A, B, C), dynamic operators by script capitals (e.g. $\mathscr{T}, \mathscr{D}, \mathscr{A}, \mathscr{B}, \mathscr{C}$) and some cost expressions by Gill capitals (e.g. $\mathbb{C}$).

The text is almost linear in that every chapter depends to a substantial degree on all those before it.

The argument in the text is largely given in theorem/proof form. This form should be regarded as neither forbidding nor pretentious, but simply as the clearest way of summarizing and punctuating a discussion. Conclusions expressed as theorems are not always either deep or novel; the point is that a definite conclusion is stated. The theorem is the string that ties up the package of a discussion.

'Exercises' are used in an almost complementary fashion: not as the 'problems' of a student text (which would inappropriate) but often as opportunities to make miscellaneous points. These are points which, although important or interesting in themselves, would have confused discussion if they had been incorporated into the main text.

PART I
LQG theory summarized

We devote Part I to a brief outline of the conventional LQG theory, in order to provide a self-contained treatment and to set the scene for subsequent generalizations. Most of the material is very standard, and the coverage correspondingly rapid. However, there is always something new to say about the commonest LQG model, and Sections 3.7 and 3.8 and Chapter 5 contain matter of some novelty, which will play an important role in the sequel.

CHAPTER 2

LQG structure: certainty equivalence

There is a part of the basic theory which is intrinsically LQG, in that it is independent of assumptions of state structure or of time homogeneity. It is with these aspects that this chapter is concerned.

2.1 HORIZONS AND HISTORIES

Exposition and treatment are both simpler if we consider optimization over a finite time interval, $h_1 \leq t \leq h_2$, say. One may well later consider either or both of the limits $h_1 \to -\infty$, $h_2 \to +\infty$, but this is indeed a later step, which can produce both simplifications and complications.

We shall, in general, argue principally in terms of the discrete time model, indicating the continuous time analogue parenthetically. Occasionally the correspondence is less evident, and the continuous case then requires separate analysis, as in Chapters 15 and 16.

The cost function will be specified in terms of the course of the process over $[h_1, h_2]$. It will imply terminal conditions, in that costs will also depend upon the terminal values of the variables. The stochastic model will define the stochastic evolution of the process for assigned values of the control variables. It must also specify initial conditions, in that it must specify sufficient statistical information about process history up to time h_1 that the optimization problem is well defined.

The current *history* of a variable will be indicated by use of the corresponding capital letter. For example, X_t will denote the history $\{x_\tau; h_1 \leq \tau \leq t\}$ of the process variable x from the initial point h_1 to the current instant t.

For present purposes we can as well take $h_1 = 0$ and set $h_2 = h$, say; the *horizon point*. The general case then easily be recovered from this. The control history over the whole optimization period will generally be written as U_{h-1} rather than as U_h, because the problem is usually formulated so as to envisage no further action at the horizon point h.

2.2 LQG STRUCTURE

The general assumptions behind a temporal optimization problem are listed in Whittle (1982). Here we note merely that a control policy essentially specifies the control u_t in terms of current observables W_t for all relevant t. Moreover, W_t will usually have the form (W_0, U_{t-1}, Y_t), where W_0 indicates initial information and U and Y indicate the control and observation histories: $U_{t-1} = (u_0, u_1, \ldots, u_{t-1})$ and $Y_t = (y_1, y_2, \ldots, y_t)$. We shall say, as a matter of definition, that the optimization problem over $[0, h]$ has *LQG structure* if the following five conditions hold:

(LQG1) For each t the control variable u_t takes values in some finite-dimensional vector space, and may take any value in this space.

(LQG2) The horizon h is fixed, and the cost function $\mathbb{C}$ whose expectation $E_\pi(\mathbb{C})$ is to be minimized can, in principle, be expressed in the reduced form $\mathbb{C} = Q(U_{h-1}, \xi)$, where U_{h-1} is the complete control path $(u_0, u_1, \ldots, u_{h-1})$ and ξ is a vector-valued noise vector.

(LQG3) The function Q is quadratic in all arguments, and positive definite in U_{h-1}.

(LQG4) Conditional on the state of information W_0 at time $t = 0$ and whatever the control policy, the noise vector ξ is normally distributed with zero mean and covariance matrix V independent of policy.

(LQG5) The observations y_t are reducible, without loss of information, to policy-independent linear functions of ξ.

Vector values are demanded, so that we may speak of linear and quadratic dependence, etc. Condition (LQG1) then requires in addition that the controls be subject to no additional constraint (such as $u \geq 0$).

The reduced form demanded in (LQG2) amounts to an elimination of the process variable. One will not wish to make this elimination for the actual optimality calculations which come later, but the proof of the certainty-equivalence principle is formally simpler for the reduced problem. For example, suppose we consider the state-structured model (1.1)–(1.5) of Section 1.1. We could solve equation (1.1) for x_t in terms of past controls and noise:

$$x_t = A^t x_0 + \sum_{j=0}^{t-1} A^j (B u_{t-j-1} + \varepsilon_{t-j})$$

$$= A^t \hat{x}_0 - A^t \Delta_0 + \sum_{j=0}^{t-1} A^j (B u_{t-j-1} + \varepsilon_{t-j}) \tag{2.1}$$

Here

$$\Delta_0 = \hat{x}_0 - x_0 \tag{2.2}$$

is the error in initial state estimate $\hat{x}_0$. If we take the noise vector ξ as having components Δ_0 and $\{\varepsilon_\tau, \eta_\tau; 1 \leq \tau \leq h\}$ then equation (2.1) expresses x_t linearly in terms of the control and noise histories U_{h-1} and ξ. Inserting this evaluation into the cost function specified by equations (1.3)–(1.5) we achieve the reduced form required in (LQG2) with the quadratic character demanded in (LQG3).

Positive definiteness is demanded in (LQG3) so that the optimal control should be well determined.

2.2 LQG STRUCTURE

The Gaussian assumption enters in (LQG4), when we demand that the only stochastic element of the model, the noise vector ξ, be normally distributed. It is also required that this noise vector be *exogenous*, i.e. generated outside the model. This manifests itself in the requirement that the properties of ξ be independent of the way in which the control values are chosen.

The final assumption requires that the observations be reducible to policy-independent linear functions of ξ, with the understanding that no information is lost in the reduction, i.e. that the original observations available at a given time can be reconstituted from the reduced observations. For example, for the model of Section 1.1 it follows in virtue of equations (1.2) and (2.1) that we can write

$$y_t = \bar{y}_t + \tilde{y}_t \tag{2.3}$$

where

$$\bar{y}_t = C\left(A^{t-1}\hat{x}_0 + \sum_0^{t-2} A^j B u_{t-j-2}\right) \tag{2.4}$$

$$\tilde{y}_t = C\left(-A^{t-1}\Delta_0 + \sum_0^{t-2} A^j \varepsilon_{t-j-1}\right) + \eta_t \tag{2.5}$$

Here $\tilde{y}_t$ is the reduced observation, indeed a linear function of ξ. It is derived from the raw observation y_t by subtracting off the effect $\bar{y}_t$ of past controls and initial prescription, calculable from observables at time t.

The same considerations hold for higher-order models. The quadratic/linear assumptions of conditions (LQG3) and (LQG5) clearly follow from the linearity of plant and observation dynamics and the quadratic dependence of $\mathbb{C}$ upon process and control variables. However, one does not need the concept of a process variable in order to characterize LQG structure. Indeed, to bring it in at this point would be confusing; the essential LQG theory is best expressed using reduced forms. We will, of course, revert later to the physically meaningful process or state description for actual optimality calculations.

These conditions permit only a marginal weakening. We demand finite dimensionality of u_t and ξ for simplicity; this could be relaxed. One could relax the positive-definiteness assumption in (LQG3) to non-negative definiteness if one were prepared to accept incompletely determined and possibly infinite controls. The assumption of a zero mean for ξ in (LQG4) is purely a matter of normalization.

Some of the LQG theory can survive in an unreduced formulation if one demands the LQ properties only of u, so allowing more general dependence of plant equation, cost function and termination rules upon the process variable; see Section 18.2.

Another possible relaxation is that the Gaussian assumption can be dispensed with if one is prepared to restrict oneself to the class of controls linear in the observations; see Theorem 2.5.3.

The true generalization (rather than weakening) that we make in Part II is to change the criterion from $E_\pi(\mathbb{C})$ to the risk-sensitive criterion (1.10). This gives us the extension from LQG structure to LEQG structure.

2.2.1 Exercise and comments

(1) Our formulation includes the possibility that that ξ can also represent the uncertainty in a 'reference signal', an exogenous variable which one wishes the controlled system to track, in some sense.

Suppose, for example, that the process vector x_t can be partitioned into sub-vectors x_{1t} and x_{2t}, satisfying

$$x_{1t} = A_1 x_{1,t-1} + B u_{t-1} + \varepsilon_{1t}$$
$$x_{2t} = A_2 x_{2,t-1} \qquad\qquad + \varepsilon_{2t}$$

and that the cost function is

$$\mathbb{C} = \sum_0^{h-1} u_t' Q u_t + \sum_0^h (x_{1t} - D x_{2t})' R_t (x_{1t} - D x_{2t})$$

with positive definite Q, R_t. The process $\{x_{2t}\}$ is then uncontrolled and, indeed, exogenous. The form of the cost function indicates that one is effectively trying to make the system variable x_{1t} follow the reference signal $D x_{2t}$. The noise vector ξ will include 'target noise' ε_{2t} as well as system noise ε_{1t} and observation noise.

2.3 THE OPTIMALITY EQUATION

The ideas of this section are independent of LQG structure or state structure. They are concerned with the recursive characterization of an optimal policy (the *dynamic programming equation*) which holds for any stochastically formulated temporal optimization problem. The treatment is set out at greater length in Whittle (1982); here we merely state the results to which we shall appeal.

Recall the notation introduced in Section 2.2, that W_t denotes the observables at time t, a listing of those quantities whose values are known at time t. We have then

$$W_t = (W_0; u_0, u_1, \ldots, u_{t-1}; y_1, y_2, \ldots, y_t) \tag{2.6}$$

That is, at time $t > 0$ one still possesses information initially available (such as the value of $\hat{x}_0$), one recalls the history of decisions U_{t-1} which have already been made, and also the observation history Y_t.

There are, of course, other quantities that it is presumed one knows, such as the values of parameters of the model (e.g. A, B, N) and clock-time t. These we take for granted: W_t represents that information which might have been different had history been rerun. As we see from (2.6), the dimension of the variable W_t

2.3 THE OPTIMALITY EQUATION

increases with t, since accumulated history is also increasing. It is a consequence of state structure that one need not carry all this history with one; see Theorem 2.3.2.

A *realizable* policy π must express the control u_t in terms of current observables W_t for all relevant t. This can be either as a definite function $u_t = u(W_t)$ or as a conditional distribution $P(u_t|W_t)$. For the case of an optimizer with complete information on action history (which we assume, by (2.6), to be the case) it is enough to restrict attention to *deterministic policies*, for which u_t shows a simple functional dependence upon W_t.

Define now

$$G(W_t) = \inf_\pi E_\pi(\mathbb{C}|W_t) \qquad (0 \le t \le h) \tag{2.7}$$

If the infimum in (2.7) can be attained for some policy then $G(W_t)$ is to be regarded as the minimal total cost that could be incurrent over $[0, h]$, conditional on information available at time t. It (or variants of it) is often termed the *value function* of the problem. If the infimum cannot be attained then it represents the lower bound which can be approached arbitrarily closely. Note that G is a different function for differing t, since the argument W_t takes values in different spaces for varying t. However, as long as we keep the t-subscript on the argument, there is no ambiguity.

Since no decisions remain at time h one may as well formally suppose that W_h implies complete information, and set

$$G(W_h) = \mathbb{C} \tag{2.8}$$

a normalization of the definition of $\mathbb{C}$.

Theorem 2.3.1. *The value function $G(W_t)$ obeys the dynamic programming or optimality equation*

$$G(W_t) = \inf_{u_t} E[\![G(W_{t+1})|W_t, u_t]\!] \qquad (0 \le t < h) \tag{2.9}$$

with terminal condition (2.8). The expectation in (2.9) is well defined in that it is independent of policy. The minimizing value of u_t in (2.9), if it exists, constitutes the optimal control: realizable in that it is a function of W_t alone; closed-loop in that it is optimal whatever the policy may have been before time t.

The theorem rests upon a number of structural assumptions, often slurred over.

Note that the optimality equation (2.9) constitutes a *backward* recursion in time. Starting from (2.8), one uses it to successively determine $G(W_{h-1})$, $G(W_{h-2}), \ldots$, in that order, and so incidentally the optimal determinations of u_{h-1}, $u_{h-2}, \ldots$.

In requiring state structure one makes the following assumptions concerning dynamics of the process variable x (now also characterized as a *state variable*), costs and observations.

(S1) *Markov dynamics*

$$P(x_t|X_{t-1}, U_{t-1}) = P(x_t|x_{t-1}, u_{t-1}) \qquad (2.10)$$

(S2) *Decomposable cost function*

$$\mathbb{C} = \sum_{t=0}^{h} \kappa_t$$

where $E(\kappa_t|X_\tau, U_\tau)$ is a function $c_t = c(x_t, u_t, t)$ of (x_t, u_t, t) alone for any $\tau \geq t$.
(S3) *State observability*: x_t is observable at time t, so that $W_t = (X_t, U_{t-1})$.

If values of control u_t which are forbidden are penalized by an infinite cost, then we can take (S2) as implying also that the set of values which u_t can adopt depends only upon x_t and t.

From these hypotheses we deduce the important and standard theorem, stating that x is indeed a state variable, in that x_t is all that one need know of observation history W_t for optimization at time t.

Theorem 2.3.2. *Under assumption* (S1)–(S3) *the infimal expected cost from time t onwards*,

$$\inf_\pi E_\pi\left(\sum_t^h c_\tau | W_t\right)$$

is a function $F_t(x_t)$ of x_t and t alone. It satisfies the optimality equation

$$F_t(x_t) = \inf_{u_t} \left(c(x_t, u_t, t) + E[F_{t+1}(x_{t+1})|x_t, u_t]\right) \qquad (t \leq h) \qquad (2.11)$$

with terminal condition

$$F_{h+1}(x_{h+1}) = 0 \qquad (2.12)$$

and the infimizing value of u yields the optimal value of u_t in closed-loop form.

The function $F_t(x)$ could be termed the 'future value function', but is generally referred to simply as the 'value function'.

2.3.1 Exercises and comments

(1) Note one way of specifying a horizon h: that $E_\pi(\mathbb{C})$ should be independent of policy after time h. Alternatively, that $E_\pi(\mathbb{C}|W_h)$ is independent of policy.
(2) To emphasize the distinction between open- and closed-loop control, consider the case of scalar x, u, with plant equation

$$x_{t+1} = x_t + u_t \qquad (2.13)$$

and cost function

$$\mathbb{C} = \sum_{0}^{h-1} u_t^2 + \Pi x_h^2$$

One finds that $F_t(x) = \Pi_t x^2$ and the optimal control is

$$u_t = -\Pi_t x_t \qquad (0 \le t < h) \tag{2.14}$$

where

$$\Pi_t = \frac{\Pi}{1 + (h-t)\Pi}$$

However, we could have regarded the problem as one of choosing $x_1, x_2, \ldots, x_h$ to minimize

$$\mathbb{C} = \sum_{0}^{h-1} (x_{t+1} - x_t)^2 + \Pi x_h^2$$

and then find the optimal path to be

$$x_t = \frac{1 + (h-t)\Pi}{1 + h\Pi} x_0$$

corresponding (cf. the plant equation) to a control

$$u_t = -\Pi_0 x_0 \tag{2.15}$$

However, control (2.14) is closed-loop; it represents the best action to be taken at time t whatever previous decisions may have been. Control (2.15) is open-loop; it is optimal only if preceding decisions have also been optimal.

The distinction becomes crucial if the process is disturbed. If the plant equation (2.13) is modified by the inclusion of a white-noise disturbing term, then the closed-loop control remains optimal. The open-loop control (2.15) does not, since it does not take account of the fact that the path of the controlled process has already deviated from prediction.

2.4 OPTIMAL ESTIMATION; LEAST SQUARES AND MAXIMUM LIKELIHOOD

As we shall see, attempts at optimal *control* imply attempts at optimal *estimation* of important unobservables from observables. This estimation amounts to *prediction* if the unobservables can be regarded as the future values of some variable. However, prediction is a special case; the essential point is simply that one must estimate an unobservable.

Let us, for the moment, forget the control context, and consider two jointly distributed random variables x and y, whose values are, respectively, unknown and known. That is, y is the observable and x the unobservable. We wish to estimate x from knowledge of y, exploiting any stochastic dependence between the variables.

For simplicity, let us suppose that x and y have been reduced to zero mean by subtraction of their mean values. One can easily undo this assumption and restore the general case; see Exercise 2.4.1(1). We shall write the covariance matrix $\text{cov}(x, y)$ between x and y as V_{xy} and can write then, in the zero-mean case:

$$\text{cov}(x, y) = V_{xy} = E(xy') \tag{2.16}$$

If this covariance matrix were zero one would say that that the random variables x and y were *mutually uncorrelated*, written $x \perp y$. The notation emphasizes that this is indeed an orthogonality relationship, in the space of random variables with inner product defined by the covariance matrix (2.16).

The simplest estimate of x in terms of y that one might consider is the *linear least square estimate* (abbreviated LSE). This is the linear estimate

$$\hat{x} = \beta y \tag{2.17}$$

where the coefficient matrix β is chosen to minimize $\hat{x} - x$ in the sense that it minimizes $E(\hat{x} - x)'M(\hat{x} - x)$ for some prescribed positive-definite matrix M.

Theorem 2.4.1. (i) *The LSE $\hat{x}$ is independent of M, and completely characterized by*

$$\hat{x} - x \perp y \tag{2.18}$$

In the case when V_{yy} is non-singular the estimate is then given explicitly by

$$\hat{x} = V_{xy} V_{yy}^{-1} y \tag{2.19}$$

and the estimation error $\Delta = \hat{x} - x$ has covariance matrix

$$V_{\Delta\Delta} = V_{xx} - V_{xy} V_{yy}^{-1} V_{yx} \tag{2.20}$$

(ii) The Gauss–Markov property. *The estimate $\hat{x}$ can also be characterized as the value of x minimizing*

$$\begin{bmatrix} x \\ y \end{bmatrix}' \begin{bmatrix} V_{xx} & V_{xy} \\ V_{yx} & V_{yy} \end{bmatrix}^{-1} \begin{bmatrix} x \\ y \end{bmatrix} \tag{2.21}$$

Thus, if x and y are jointly normally distributed, then $\hat{x}$ can also be characterized both as the maximum likelihood estimate (MLE) of x and as the conditional expectation $E(x|y)$, with conditional covariance matrix (2.20).

For proof, see Appendix 3. For the moment, we simply note two points. First, the 'geometric' interpretation of (2.18): this relation characterizes $\hat{x}$ as the statistical projection of x upon y in that the difference $\hat{x} - x$ is required to be orthogonal to y. Second, the dual extremal characterizations of $\hat{x}$: as LSE or MLE.

Despite the fact that both extremal criteria yield the same estimate, one can be much more convenient than the other, and we shall find the ML characteriza-

2.4 OPTIMAL ESTIMATION; LEAST SQUARES AND MAXIMUM LIKELIHOOD

tion the natural one to employ in this work. For example, the classical characterization of the Wiener predictor is as an LSE, but there are very great advantages in taking the ML approach instead; see Sections 1.4 and Exercise 5.3.1.

In order to emphasize the dependence of $\hat{x}$ upon the explanatory variable(s) y we shall write the estimate as

$$\hat{x} = \mathscr{E}(x|y) \qquad (2.22)$$

The operator $\mathscr{E}$ is indeed a projection operator, expression (2.22) denoting the projection of x upon y, with the covariance defining the inner product. It has, for example, the properties characteristic of a projection:

$$\mathscr{E}(\mathscr{E}(x|y)|y) = \mathscr{E}(x|y) \qquad (2.23)$$

and

$$\mathscr{E}(x|y_1, y_2) = \mathscr{E}(x|y_1) + \mathscr{E}(x|y_2) \qquad (2.24)$$

if $y_1 \perp y_2$.

In cases of joint normality we can replace $\mathscr{E}$ by the conventional expectation operator E, as asserted in Theorem 2.4.1(ii).

If $\{y_1, y_2, \ldots\}$ is an observation sequence then

$$\zeta_t = y_t - \mathscr{E}(y_t|Y_{t-1}) \qquad (2.25)$$

is termed the *innovation* in the y-sequence at time t: that part of y_t which is unpredictable from previous observation history.

Theorem 2.4.2. *The innovations sequence is orthogonal.*

By this we mean that it is self-orthogonal: i.e. that $\zeta_s \perp \zeta_t$ for $s \neq t$.

Proof. We have $\zeta_t \perp Y_{t-1}$, by equation (2.18). However, the innovations $\zeta_1, \zeta_2, \ldots, \zeta_{t-1}$ are linear functions of Y_{t-1}, and so are also orthogonal to ζ_t. □

The estimate $\mathscr{E}(x|Y_t)$ of x in terms of observation history can equally well be regarded as an estimate in terms of innovations history, with the advantage that

$$\mathscr{E}(x|Y_t) = \sum_{j=1}^{t} b_j \zeta_j \qquad (2.26)$$

where the b_j are independent of t. This follows from the orthogonality of the innovations and equation (2.24).

The change in estimate as t is varied can thus be written

$$\chi_{t+1} = \mathscr{E}(x|Y_{t+1}) - \mathscr{E}(x|Y_t) = b_{t+1}\zeta_{t+1} \qquad (2.27)$$

whence we derive a result to which we shall soon make appeal:

Corollary 2.4.3

$$\mathscr{E}(\chi_{t+1}|Y_t) = 0$$

Let us now revert to the control optimization problem in the reduced LQG form formulated in Section 2.2: the minimization of $E_\pi(\mathbb{C}) = E_\pi Q(U_{h-1}, \xi)$ with u_t restricted to being a function of $W_t = (W_0, Y_t, U_{t-1})$. We shall in fact suppress W_0, by assuming all statements made conditional on W_0. That is, random variables whose values are specified by W_0 are regarded as constants. Let us define

$$\xi^{(t)} = \mathscr{E}(\xi|\tilde{Y}_t) \tag{2.28}$$

where the $\tilde{y}_t$ are the reduced observations, reduced to being linear functions of the total noise vector ξ.

Theorem 2.4.4. (i) *If the elements of ξ are normally distributed then*

$$E_\pi(\xi|W_t) = \xi^{(t)} \tag{2.29}$$

(ii) *If one does not make the assumption of normality, but assumes that u_t is a linear function of Y_t for all relevant t, then*

$$\mathscr{E}(\xi|W_t) = \xi^{(t)} \tag{2.30}$$

Proof. We have

$$E_\pi(\xi|W_t) = E(\xi|\tilde{Y}_t, U_{t-1}) = E(\xi|\tilde{Y}_t)$$

the second equality following since U_{t-1} is a function of $\tilde{Y}_{t-1}$. However, in case (i), we can then identify $E(\xi|\tilde{Y}_t)$ with $\xi^{(t)}$, in virtue of Theorem 2.4.1(ii).

In case (ii) we can simply note that an estimate linear in $W_t = (\tilde{Y}_t, U_{t-1})$ is linear in $\tilde{Y}_t$, whence assertion (2.30) follows. □

2.4.1 Exercises and comments

(1) If means are not assumed to be zero then the linear estimate (2.17) should be generalized to the non-homogeneous form

$$\hat{x} = \alpha + \beta y$$

In the cases when x and y in fact have zero means one finds that the characterization of $\hat{x}$ either an LSE or as a MLE requires that $\alpha = 0$. In the more general case α is then effectively determined by the relation

$$\hat{x} - E(x) = \beta(y - E(y))$$

This is equivalent to saying that α must be such as to give $\hat{x} - x$ zero expectation. That is, the LSE can now be characterized as the linear function $\hat{x}$ of y satisfying

$\hat{x} - x \perp (1, y)$, where the inner product between two random variables x and y which defines orthogonality is now the mean product $E(xy')$ rather than the covariance $\text{cov}(x, y)$.

(2) Verify equations (2.23) and (2.24).

(3) Note the evaluation of b_j in (2.26):

$$b_j = \text{cov}(x, \zeta_j) \text{cov}(\zeta_j)^{-1}$$

(4) If x, y are jointly Gaussian, with zero means, then their joint probability density is of the form

$$f(x, y) \propto \exp[-\tfrac{1}{2}(x'I_{xx}x + x'I_{xy}y + y'I_{yx}x + y'I_{yy}y)]$$

where

$$\begin{bmatrix} I_{xx} & I_{xy} \\ I_{yx} & I_{yy} \end{bmatrix} = \begin{bmatrix} V_{xx} & V_{xy} \\ V_{yx} & V_{yy} \end{bmatrix}^{-1}$$

The MLE of x for observed y is then

$$\hat{x} = -I_{xx}^{-1} I_{xy} y$$

and the covariance of x conditional on the value of y is I_{xx}^{-1}. Since, by the Gauss–Markov theorem, these quantities are to be identified with expressions (2.19) and (2.20), we deduce the identities between components of the mutually reciprocal covariance and information matrices:

$$V_{xy} V_{yy}^{-1} = -I_{xx}^{-1} I_{xy}$$

$$V_{xx} - V_{xy} V_{yy}^{-1} V_{yx} = I_{xx}^{-1}$$

Note that V_{xy} will not change if we move from discussing the joint distribution of (x, y) to discussing that of an extended set (x, y, z), whereas I_{xy}, in general, will.

2.5 THE CERTAINTY-EQUIVALENCE PRINCIPLE (CEP)

Let us now specify LQG structure by making all the reduced-form assumptions of Section 2.2. As time moves forward one gains more and more information on the essential unknown (or random variable) of the problem: the noise vector ξ. Suppose, however, that one had perfect information from the beginning, in that the value of ξ was known perfectly already at $t = 0$. One would then determine the optimal values of $u_t, u_{t+1}, \ldots, u_{h-1}$ simply by minimizing $Q(U_{h-1}, \xi)$ with respect to these variables, so obtaining a set of linear relations between these variables and the known quantities U_{t-1} and ξ. Indeed, if we define

$$Q_t(U_{t-1}, \xi) = \min_{u_t, u_{t+1}, \ldots, u_{h-1}} Q(U_{h-1}, \xi) \qquad (2.31)$$

(also a quadratic form in its arguments) then we can make the identification

$$G(W_t) = Q_t(U_{t-1}, \xi) \qquad (2.32)$$

and the optimality equation (2.9) reduces simply to

$$Q_t(U_{t-1}, \xi) = \min_{u_t} Q_{t+1}(U_t, \xi) \qquad (2.33)$$

Equation (2.33) is indeed an immediate consequence of definition (2.31), but it *is* the optimality equation in this degenerate case. Moreover, the minimizing value of u_t is the optimal value of u_t in its closed-loop form; it is optimal, however previous controls U_{t-1} may have been chosen.

It is one of the most important consequences of LQG structure that these conclusions have a close analogue in the general case: that of imperfect observation.

Theorem 2.5.1. The certainty-equivalence principle (CEP). *Assume conditions (LQG1–5) of Section 2.2, and suppose policy π to be chosen to minimize $E_\pi(\mathbb{C})$. Then*
(i) *The optimal value of u_t, in closed-loop form, is that minimizing $Q_{t+1}(U_t, \xi^{(t)})$, where $\xi^{(t)}$ is the estimate $\mathscr{E}(\xi | W_t) = \mathscr{E}(\xi | \tilde{Y}_t)$.*
(ii) *The value function $G(W_t)$ of the problem has the evaluation*

$$G(W_t) = Q_t(U_{t-1}, \xi^{(t)}) + \cdots \qquad (2.34)$$

where $+\cdots$ denotes policy-independent terms.

That is, roughly speaking, the conclusions of the perfect-information case carry over, except that at time t we replace ξ by its current LSE $\xi^{(t)}$. This is just what is meant by 'certainty equivalence'.

Proof. Relation (2.34) certainly holds at $t = h$, since

$$G(W_h) = Q(U_{h-1}, \xi)$$

if we make the convention (which cannot affect past actions) that W_h conveys full information. Assume then that (2.34) holds at $t + 1$. The optimality equation then becomes

$$G(W_t) = \inf_\pi E_\pi[Q_{t+1}(U_t, \xi^{(t+1)}) | W_t] + \cdots$$

$$= \inf_\pi E_\pi[Q_{t+1}(U_t, \xi^{(t)} + \chi_{t+1}) | W_t] + \cdots \qquad (2.35)$$

where χ_{t+1} is the change in estimate

$$\chi_{t+1} = \xi^{(t+1)} - \xi^{(t)}$$

Now, by equation (2.27) and the identification (2.29) we have

$$E_\pi(\chi_{t+1} | W_t) = 0$$

Further, by the policy independence of the statistics of ξ and so of $\tilde{Y}_t$, it follows that $E_\pi(\chi_{t+1} \chi'_{t+1} | W_t)$ is policy independent. If we expand expression (2.35) in

2.5 THE CERTAINTY-EQUIVALENCE PRINCIPLE (CEP)

powers of χ_{t+1} we thus derive zero contribution from the first-order terms, a policy-independent contribution from the second-order terms, and there are no terms of higher order, in virtue of the quadratic nature of Q. Equation (2.35) thus yields

$$G(W_t) = \inf_\pi E_\pi[Q_{t+1}(U_t, \xi^{(t)})|W_t] + \cdots$$
$$\geq \min_{u_t} Q_{t+1}(U_t, \xi^{(t)}) + \cdots$$
$$= Q_t(U_{t-1}, \xi^{(t)}) + \cdots \quad (2.36)$$

However, the inequality in (2.36) becomes an equality if π is such that u_t is indeed given the value minimizing $Q_{t+1}(U_t, \xi^{(t)})$. This choice is admissible, since it makes u_t a function of W_t. Both assertions of the theorem then follow. □

The characterization of the conclusion as 'certainty equivalence' is apposite in that one optimizes exactly as one would in the full-information case, except that the noise vector ξ is replaced by its current best estimate $\xi^{(t)}$. A less condensed statement of the principle is the following.

Theorem 2.5.2. (i) *The optimal value u_t° of u_t is determined by minimizing $Q(U_{h-1}, \xi^{(t)})$ with respect to remaining decisions $u_t, u_{t+1}, \ldots, u_{h-1}$.*
(ii) *Let $u_\tau^{(t)}$ denote the value of u_τ thus determined. Then*

$$u_\tau^{(t)} = \begin{cases} u_\tau & \tau < t \\ u_t^\circ & \tau = t \\ E(u_\tau^\circ|W_t) & \tau > t \end{cases} \quad (2.37)$$

Proof. Statement (i) is nothing but a restatement of Theorem 2.5.1(i). Relation (2.37) is trivial for $\tau < t$ and follows from (i) for $\tau = t$. It follows for $\tau > t$ from the fact that $u_\tau^\circ - u_\tau^{(t)}$ is then a linear function of the innovations $\zeta_{t+1}, \zeta_{t+2}, \ldots, \zeta_\tau$ of the reduced observation series $\{\tilde{y}_t\}$. □

The advantage of the statement of the CEP in the form (i) is that it avoids the rather telescoped formulation associated with a dynamic programming argument, and sets the problem out as an optimization over future plans. The values $u_\tau^{(t)}$ ($\tau > t$) of future decisions determined in this way will not, in general, actually be optimal. However, in the sense of assertion (ii) they do constitute the best estimate one can make at time t of what future optimal decisions will be. This view of the optimization is advantageous: that the calculation of the optimal value of u_t at time t is embedded in the determination of a *provisional forward plan* of future actions. This consciousness of future actions tends to be buried in a dynamic programming treatment, but is certainly very much the way in which the intuitive optimizations of daily life are made. It is then reassuring to have a mathematical treatment in which these appear.

There is also a second-order version of the CEP.

Theorem 2.5.3. *Restricted certainty equivalence. The assertions of Theorem 2.5.1 still hold if, in place of the Gaussian assumption of condition (LQG4), one requires that controls be linear in observables.*

The proof is analogous to that of Theorem 2.5.1, with appeal to the second rather than the first part of Theorem 2.4.4.

The CEP is sometimes called the *separation principle*, the separation being that of control and estimation. Indeed, the estimate $\xi^{(t)}$ is independent of control, in that its representation in terms of the reduced observations $\tilde{Y}_t$ is policy independent. Control could also be said to be independent of estimation, in the weaker sense that one derives the optimal control simply by the substitution of $\xi^{(t)}$ for ξ in the full-information optimal control. However, there is another characterization of the separation principle, and it is only this which survives as we move into the LEQG case; see Theorem 7.2.1.

2.5.1 Exercises and comments

(1) Suppose that $Q_t(U_{t-1}, \xi) = \xi' R_t \xi$ plus terms of lower order in ξ, and define $M_{t+1} = E(\chi_{t+1} \chi'_{t+1} | \tilde{Y}_t)$. Then we can identify the policy-independent term in $G(W_t)$ as

$$G(W_t) - Q_t(U_{t-1}, \xi^{(t)}) = \sum_{t+1}^{h} \text{tr}(M_\tau R_\tau)$$

(2) The requirement that either the injected noise be Gaussian or the controls be linear in observables is necessary to ensure the relation $\chi_{t+1} \perp U_t$, to which appeal is made in the proof of the CEP.

To demonstrate necessity, consider the one-stage scalar problem of choosing u to minimize $E(x - u)^2$, given that one knows the value of a variable y jointly random with x. If x itself were known then the minimizing value would be $u = x$. If x is not known then certainty equivalence would require the minimizing value to be $u = \hat{x} = \mathscr{E}(x|y)$. This is indeed the minimizing value if u is required to be linear in y. The unrestricted minimizing value is $u = E(x|y)$, which coincides with $\mathscr{E}(x|y)$ when x and y are jointly normal, but not, in general, otherwise.

2.6 CONTROL/OPTIMIZATION DUALITY: THE UNREDUCED CASE

The consecutive construction of the estimates $\xi^{(t)}$ as time moves forward can be given a formulation parallel to that of the consecutive construction of the full-information optimal controls u_t as time moves back. In fact, suppose that the joint density of the reduced observations $\tilde{Y}_h$ is written

$$f(\tilde{Y}_h) \propto \exp[-\tfrac{1}{2}Q^*(\tilde{Y}_h)]$$

where Q^* is also a quadratic form. Then the estimates $\tilde{y}_\tau^{(t)}$ of future reduced observations $\tilde{y}_\tau$ at time t are obtained by minimizing $Q^*(\tilde{Y}_h)$ with respect to $\tilde{y}_\tau$ ($\tau > t$). From these estimates and the observed values $\tilde{y}_\tau$ ($\tau \leq t$) the estimate $\xi^{(t)}$ can be constructed, since we assume that ξ is determinable from $\tilde{Y}_h$.

This partial minimization of the quadratic form Q^* is directly analogous to the partial minimization (2.31) of the form Q which determined the optimal full-information estimates. This indicates a kind of duality between optimal control and optimal estimation.

In fact, all the reductions and eliminations which we have made for ease of proof will not be made in actual cases, and the situation will rather be as follows. The cost function will be written as a quadratic function of the path of the process

$$\mathbb{C} = \mathbb{C}(X_h, U_{h-1}) \tag{2.38}$$

Similarly, we can write down the joint density of the process variables and the observations for given values of the control variables as

$$f(X_h, Y_h | U_{h-1}) \propto \exp[\![-\tfrac{1}{2}\mathbb{D}(X_h, Y_h, U_{h-1})]\!] \tag{2.39}$$

where $\mathbb{D}$ is also a quadratic function of the arguments indicated. (Note that no control policy has been introduced as yet, and the u_t are not defined as random variables. The control history U_{h-1} *parametrizes* distribution (2.39) rather than conditions it.

Distinguish now between decisions as yet unmade (controls as yet undetermined) at a given time t and unobservables at time t (such as future values of x and y, and also past values of x if these have not been observed). One then estimates all unobservables at time t by minimizing $\mathbb{D}$ with respect to these unobservables. The burden of the CEP is that one can derive the optimal control u_t° by substituting these estimates of unobservables in $\mathbb{C}$, and then minimizing the consequent quadratic form with respect to all decisions (control variables) as yet undetermined. This is the view of the procedure which will prove natural when we take the more general LEQG formulation, and which we shall formally prove for that case.

2.7 NOTES ON THE LITERATURE

The first statement and proof of the certainty-equivalence principle seems to be due to Theil (1957). The treatment given here, which relates the principle more closely to dynamic programming ideas, is that given in Whittle (1982).

CHAPTER 3

The Markov case: control rules

Most of the material of this chapter is very standard. However, we set it out systematically so as to have a point of reference for the generalizations of the remaining parts of the book. It also serves as a point of reference in that, in the author's view, the analysis begins to show some streamlining and style *first* in the more general formulations, foreshadowed in Section 3.7 and continued in Parts II and III. Some of the material of Section 3.8 is new, and plays an essential role in the sequel.

We derive control rules on the assumption of perfect state information. The rules for the case of imperfect observation then follow by certainty equivalence, under appropriate hypotheses.

3.1 SOME POINTS FOR STATE-STRUCTURED PROCESSES

If we make the state-structure (or Markov) assumptions (S1–3) of Section 2.3, and assume in fact that the cost function has the specific form

$$\mathbb{C} = \sum_{t=0}^{h-1} c(x_t, u_t, t) + \mathbb{C}_h(x_h) \tag{3.1}$$

then the optimality equation (2.11) becomes

$$F_t(x_t) = \inf_{u_t} \left(c(x_t, u_t, t) + E[F_{t+1}(x_{t+1}) | x_t, u_t] \right) \qquad (t < h) \tag{3.2}$$

with terminal condition

$$F_h(x) = \mathbb{C}_h(x) \tag{3.3}$$

Here $F_t(x)$ is the future *value function*, the infimal value of future cost at time t if $x_t = x$.

Suppose the process is time homogeneous in that neither the instantaneous cost c nor the conditional expectation $E[\cdot | x_t, u_t]$ depend explicitly on t. Then $F_t(x)$ will depend upon t and horizon h only in the combination $s = h - t$, the *time-to-go*. Let us exhibit this by writing it also as $F_{(s)}(x)$. It is then often convenient to write the optimality equation and its terminal condition in the condensed forms

$$F_{(s)} = \mathscr{L} F_{(s-1)} \qquad (s > 0) \tag{3.4}$$

$$F_{(0)} = \mathbb{C}_h \tag{3.5}$$

Here $\mathscr{L}$ is an operator converting functions of x to functions of x by the rule

$$(\mathscr{L}\phi)(x) = \inf_u (c(x, u) + E[\phi(x_{t+1})|x_t = x, u_t = u]) \tag{3.6}$$

and we have suppressed the x-arguments in equations (3.4) and (3.5). Let us denote the class of scalar functions of state by $\mathscr{F}$; the operator $\mathscr{L}$ then maps $\mathscr{F}$ onto itself. We can term $\mathscr{L}$ the *one-step optimization operator*; expression (3.6) is the minimal cost incurred from state x if $\phi(x)$ is the cost incurred from x after one further step. The following properties of $\mathscr{L}$ are readily proved.

Theorem 3.1.2. *The operator $\mathscr{L}$ is monotone in that if $\phi \geq \psi$ for ϕ, ψ in $\mathscr{F}$ then $\mathscr{L}\phi \geq \mathscr{L}\psi$. This has as consequence that if $F_{(1)} \geq F_{(0)}$ ($F_{(1)} \leq F_{(0)}$) then the sequence $\{F_{(s)}\}$ is monotone non-decreasing (non-increasing).*

A point of the formulation in terms of time-to-go is that one may expect that, under reasonable conditions, the optimal control will become time independent in the infinite-horizon limit; i.e. as $h \to \infty$ or $s \to \infty$.

3.2 THE LQG REGULATION MODEL

Consider the following combination of LQG and state-structure assumptions. It essentially specifies a model for optimization of regulation to the set point ($x = 0, u = 0$), with perfect state observation, and without any kind of disturbance, either deterministic or random.

(M1) The state variable x and control variable u take values in R^n and R^m, respectively, and may take any values in these spaces.

(M2) The plant equation has the linear deterministic form

$$x_t = Ax_{t-1} + Bu_{t-1} \tag{3.7}$$

(M3) Operation is over the fixed time interval $0 \leq t \leq h$ with quadratic cost function

$$\mathbb{C} = \sum_{t=0}^{h-1} c(x_t, u_t) + \mathbb{C}_h(x_h) \tag{3.8}$$

where

$$c(x, u) = x'Rx + x'S'u + u'Sx + u'Qu = \begin{bmatrix} x \\ u \end{bmatrix}' \begin{bmatrix} R & S' \\ S & Q \end{bmatrix} \begin{bmatrix} x \\ u \end{bmatrix} \tag{3.9}$$

$$\mathbb{C}_h(x) = x'\Pi_h x \tag{3.10}$$

and $\begin{bmatrix} R & S' \\ S & Q \end{bmatrix} \geq 0, Q > 0, \Pi_h \geq 0.$

(M4) The current value of state is always observable.

3.2 THE LQG REGULATION MODEL

The positive-definiteness hypotheses of (M3) imply that deviations of x and u from zero are penalized. The model gives us the required starting point, although we shall later wish to relax all of assumptions (M1–4).

Theorem 3.2.1. *Assume conditions* (M1–4). *Then the value function has the quadratic form*

$$F_t(x) = x'\Pi_t x \qquad (t \leq h) \qquad (3.11)$$

and the optimal control the linear form

$$u_t = K_t x_t \qquad (t \leq h) \qquad (3.12)$$

Here the matrices Π_t are determined by the backward recursion

$$\Pi_t = f\Pi_{t+1} \qquad (t < h) \qquad (3.13)$$

where f is an operator having the action

$$f\Pi = R + A'\Pi A - (S' + A'\Pi B)(Q + B'\Pi B)^{-1}(S + B'\Pi A) \qquad (3.14)$$

and Π_h has the prescribed value implied by (3.10). *The matrix K_t has the evaluation*

$$K_t = -(Q + B'\Pi_{t+1}B)^{-1}(S + B'\Pi_{t+1}A) \qquad (t < h) \qquad (3.15)$$

Proof. This is by backwards induction. By assumption (3.10) the form (3.11) holds at $t = h$, and $\Pi_h \geq 0$. Assume, then, that it holds at $t + 1$ with $\Pi_{t+1} \geq 0$. The optimality equation equation (3.2) then becomes

$$F_t(x) = \inf_u \left(c(x, u) + (Ax + Bu)'\Pi_{t+1}(Ax + Bu) \right) \qquad (3.16)$$

where $c(x, u)$ is given by (3.9). The total bracketed expression on the right is a non-negative quadratic form in x and u, positive definite in u. The assertions (3.11)–(3.15) then follow from standard results concerning the minimization of quadratic forms; see Exercise 3.2.1(1). □

The theorem represents a very considerable reduction, in that the linear/quadratic character of the assumptions propagate through to the forms of the optimal control and value function. Complete solution of the problem reduces to determination of the time-dependent matrices Π_t and K_t, and this, as we see, reduces to solution of the recursion (3.13).

Recursion (3.13), with f defined as in (3.14), is the celebrated *Riccati equation*, whose character and solution are so fundamental to the whole study. It may not appear particularly natural or attractive at first sight, but has in fact a clear rationale. The equation has a quasilinear character (Exercise 3.2.3), looks neater in the more general context of path-integral extremization (Section 11.1) and has an immediate relation to the most elegant and powerful methods of all: those of operator factorization (Section 12.3).

The plant equation for the optimally controlled process reduces, in virtue of (3.7) and (3.12), to the simple linear recursion

$$x_t = \Gamma_{t-1} x_{t-1} \tag{3.17}$$

where

$$\Gamma_t = A + BK_t \tag{3.18}$$

The matrix Γ_t is often called the *gain matrix*, although this is one of those terms which are used inconsistently in a number of contexts.

As the time-to-go increases (e.g. if we consider fixed t and let the horizon point h tend to infinity) one might expect that the matrix Π_t would tend to a limit Π, the quadratic form $x'\Pi x$ then representing the minimal cost of taking the state variable to zero from an initial value x, with no time limitation. The matrices K_t and Γ_t would then tend to corresponding limits K and Γ. The optimally controlled plant equation (3.17) will then follow the path

$$x_t = \Gamma^t x_0 \tag{3.19}$$

If the optimally controlled system is indeed successful in taking the state variable to zero from any initial value x_0, then we see from (3.19) that Γ must be a *stability matrix* (i.e. such that $\Gamma^t \to 0$ as $t \to +\infty$). One will expect this stable behaviour if $x_t \to 0$ is necessary for finite cost and possible at finite cost.

If these infinite-horizon limits exist we shall speak of the problem as being *horizon stable*. We return to these matters in Section 3.4.

3.2.1 Exercises and comments

(1) Consider a quadratic form in vectors x and u

$$\begin{bmatrix} x \\ u \end{bmatrix}' \begin{bmatrix} \Pi_{xx} & \Pi_{xu} \\ \Pi_{ux} & \Pi_{uu} \end{bmatrix} \begin{bmatrix} x \\ u \end{bmatrix}$$

the matrix of this form being symmetric and non-negative definite, with $\Pi_{uu} > 0$. Suppose the form is to be minimized with respect to u. Show that the minimizing value is $u = -\Pi_{uu}^{-1} \Pi_{ux} x$, and that the minimal value is $x'(\Pi_{xx} - \Pi_{xu} \Pi_{uu}^{-1} \Pi_{ux}) x$, a non-negative definite quadratic form in x.

(2) Note that the matrix S of the cross-term can be normalized to zero by a change of the control variable to $u_t^* = u_t - Q^{-1} S x_t$. This implies the simultaneous transformations $A \to A^* = A - BQ^{-1}S$ and $R \to R^* = R - S'Q^{-1}S$.

(3) The left-hand member of (3.16) is $x' \Pi_t x$, defining Π_t. The consequent identity in quadratic forms could be written

$$\Pi_t = \inf_K (R + K'S + S'K + K'QK + (A + BK)' \Pi_{t+1} (A + BK)) \tag{3.20}$$

Here K is a matrix of appropriate size ($n \times m$) and the infimum is understood in the ordering of positive definiteness. Equation (3.20) can be regarded as the Riccati equation exhibited in quasilinear form: representing Π_t as an extreme of a class of linear functions of Π_{t+1}.

(4) Consider a stationary policy $u_t = Kx_t$, where K is not necessarily chosen optimally. This will have h-horizon cost $x_0' P_{(h)} x_0$, where

$$P_{(h)} = \sum_{t=0}^{h-1} (\Gamma')^t c(K) \Gamma^t + (\Gamma')^h \Pi_h \Gamma^h \qquad (3.21)$$

and $\Gamma = A + BK$, $c(K) = R + S'K + K'S + K'QK$. If Γ is a stability matrix (so that $u = Kx$ is a stabilizing control) then expression (3.21) has a limit value as h increases (if the terminal cost matrix Π_h is held constant, say). Note that this limit matrix P satisfies the equation

$$P - \Gamma' P \Gamma = c(K) \qquad (3.22)$$

(5) Suppose the plant equation disturbed by white noise

$$x_t = Ax_{t-1} + Bu_{t-1} + \varepsilon_t$$

of covariance matrix N. The optimal estimate of future noise is zero, so, by the certainty-equivalence principle, the presence of noise will not affect the optimal control in its feedback form: (equations (3.12) and (3.15)). Show that the value function has the form

$$F_t(x) = x' \Pi_t x + \gamma_t$$

where Π_t has the same value as in the noise-free case and

$$\gamma_t - \gamma_{t+1} = \text{tr}(N \Pi_{t+1})$$

3.3 LQG REGULATION WITH DISTURBANCES AND VARIABLE SET POINTS

Suppose the plant equation (3.7) modified to

$$x_t = Ax_{t-1} + Bu_{t-1} + \alpha_t \qquad (3.23)$$

where $\{\alpha_t\}$ is a *known* (and so deterministic) sequence of disturbances.

Theorem 3.3.1. *Assume conditions (M1–4) of Section 3.2, with the sole modification that the plant equation takes the disturbed form (3.23). Then the value function has the form*

$$F_t(x) = x' \Pi_t x - 2\sigma_t' x + \gamma_t \qquad (3.24)$$

where Π_t has the same evaluation as in Theorem 3.2.1 and σ_t satisfies the backward recursion

$$\sigma_t = \Gamma_t'(\sigma_{t+1} - \Pi_{t+1} \alpha_{t+1}) \qquad (t < h) \qquad (3.25)$$

The corresponding optimal control is

$$u_t = K_t x_t + (Q + B'\Pi_{t+1}B)^{-1}B'(\sigma_{t+1} - \Pi_{t+1}\alpha_{t+1})$$

$$= K_t x_t - (Q + B'\Pi_{t+1}B)^{-1}B'\left(\sum_{j=0}^{h-t-1}\Gamma'_{t+1}\Gamma'_{t+2}\cdots\Gamma'_{t+j}\Pi_{t+j+1}\alpha_{t+j+1}\right.$$

$$\left. - \Gamma'_{t+1}\cdots\Gamma'_{h-1}\sigma_h\right) \qquad (t < h) \tag{3.26}$$

Here K_t has the previous evaluation (3.15).

The proof follows the inductive pattern of that for Theorem 3.2.1. The optimal control (3.26) is, as ever, in closed-loop form, and so optimal whatever policy has been before time t. Equation (3.26) represents the control in *feedback/feedforward* form. That is, it contains the familiar feedback term $K_t x_t$, which induces the system to correct existing state deviations. However, it also contains a linear form in future disturbances, which induces the system to anticipate these disturbances in an economic way. (Note that, with the assumed terminal cost function (3.10), we would have $\sigma_h = 0$. However, we shall later need the more general case.)

It is interesting that the dynamics of feedforward are, in a sense, adjoint to those of feedback, in that they involve the matrices Γ'_t, where Γ_t is the gain matrix of the optimally controlled undisturbed system (3.17). The mutual adjointness is, in fact, deep rooted and complete; see Sections 3.7, 10.4 and 10.6.

The modified problem is no longer quite time homogeneous, because the injected sequence $\{\alpha_t\}$ is time dependent, However, if the undisturbed system is horizon stable and if $\sum_0^\infty (\Gamma')^t \alpha_t$ is absolutely convergent then the optimal control will incorporate this input in a stationary (i.e. translation-invariant) manner. Indeed, the obvious limit version of relation (3.26) is

$$u_t = Kx_t - (Q + B'\Pi B)^{-1}B'\sum_{j=0}^\infty (\Gamma')^j \Pi \alpha_{t+j+1} \tag{3.27}$$

Expression (3.27) and the stability of Γ imply that optimal feedforward produces an automatic discounting of the future, in that the coefficient of α_{t+j} converges exponentially fast to zero as j increases. This reflects, of course, the fact that distant contingencies are taken less seriously, because there is more time to deal with them.

We did not evaluate the absolute term γ_t of expression (3.24). This term is of interest in that it reflects the cost of future operations, but it does not reflect the sensitivity of this cost to current state value or affect the choice of control. However, we do evaluate it in Section 11.1 for a general path-integral formulation of the problem, for which all formulae are much more compact.

Theorem 3.3.1 has two useful corollaries. One concerns the case when $\{\alpha_t\}$ is random. More properly expressed, one does not have observations which deter-

3.3 LQG REGULATION WITH DISTURBANCES AND VARIABLE SET POINTS 37

mine the future values of α, and so is compelled to assume these random, with a model which may allow some degree of effective prediction. If one assumes conditions that allow application of the CEP then the optimal course is to use the control rule (3.26) with α_τ replaced by its ML predictor $\alpha_\tau^{(t)}$.

Corollary 3.3.2. *Suppose that the sequence of disturbances $\{\alpha_t\}$ has an unknown future course, but that it is generated by a linear policy-independent process of which one has linear observations (in addition to knowledge of state and control histories). Suppose also that either $\{\alpha_t\}$ and observation noise are jointly Gaussian, or that controls are required to be linear in observables. Then the evaluations (3.24) and (3.26) of value function and optimal control remain valid if the ML predictor $\alpha_\tau^{(t)}$ is substituted for α_τ ($\tau > t$).*

As stated, the proof is simply by appeal to the CEP (Theorems 2.5.1 and 2.5.3). By a 'linear process' we mean that the variable α_t obeys linear noise-driven dynamics; by 'linear observation' we mean that we observe linear (possibly noise-corrupted) functions of the α-process; by 'policy-independent' we mean that the α-process is exogenous (and so unaffected by control) and the observations on it also unaffected by control decisions.

Imperfect observation of the α-process will indeed modify the value function in that it will increase the absolute term γ_t. However, the x-dependent part of the value function is affected only in the way indicated. Note that the supposition that X_t is observable at time t implies that the values of $\alpha_t, \alpha_{t-1}, \ldots$ are then also observable, since they are deducible from the plant equation (3.23).

An extreme case is that in which α_t is white noise with an unknown future, when we simply have $\alpha_\tau^{(t)} = 0$ ($\tau > t$).

The other corollary covers the case of variable set points by simple appeal to Theorem 3.3.1.

Corollary 3.3.3. *Assume conditions (M1–4) of Section 3.2 modified in that the plant equation has the disturbed form (3.23) and the cost function is modified to*

$$\mathbb{C} = \sum_{t=0}^{h-1} \begin{bmatrix} x - \bar{x} \\ u - \bar{u} \end{bmatrix}_t' \begin{bmatrix} R & S' \\ S & Q \end{bmatrix} \begin{bmatrix} x - \bar{x} \\ u - \bar{u} \end{bmatrix}_t + (x - \bar{x})_h' \Pi_h (x - \bar{x})_h \qquad (3.28)$$

where the sequences $\{\alpha_t\}$, $\{\bar{x}_t\}$ and $\{\bar{u}_t\}$ are all known. Then

$$F_t(x) = (x - \bar{x}_t)' \Pi_t (x - \bar{x}_t) - 2\sigma_t'(x - \bar{x}_t) + \gamma_t \qquad (3.29)$$

and the optimal control is

$$u_t = \bar{u}_t + K_t x_t - (Q + B' \Pi_{t+1} B)^{-1} B'$$
$$\times \left(\sum_{j=0}^{h-t-1} \Gamma'_{t+1} \Gamma'_{t+2} \ldots \Gamma'_{t+j} \Pi_{t+j+1} (\alpha_{t+j+1} - \bar{\alpha}_{t+j+1}) - \Gamma'_{t+1} \ldots \Gamma'_{h-1} \Pi_h \bar{x}_h \right) \qquad (3.30)$$

Here Π_t and σ_t have the same evaluations as in Theorem 3.3.1, except that α_t is replaced by $\alpha_t - \bar{\alpha}_t$, where

$$\bar{\alpha}_t = \bar{x}_t - A\bar{x}_{t-1} - B\bar{u}_{t-1} \tag{3.31}$$

The problem formulated is now one of optimal regulation to a known path $\{\bar{x}_t, \bar{u}_t\}$ with a deterministically disturbed plant equation.

Proof. If one changes indeed to new variables $x^* = x - \bar{x}$ and $u^* = u - \bar{u}$ then one has effectively normalized $\bar{x}$ and $\bar{u}$ to zero and has a transformed plant equation

$$x_t^* = Ax_{t-1}^* + Bu_{t-1}^* + \alpha_t^*$$

with $\alpha_t^* = \alpha_t - \bar{\alpha}_t$. Conclusions then follow by application of Theorem 3.3.1 to the normalized problem. □

3.3.1 Exercise and comments

(1) Note that the stationary form of the optimally controlled plant equation is

$$(I - \Gamma\mathcal{T})x_t = B(Q + B'\Pi B)^{-1}B'(I - \Gamma'\mathcal{T}^{-1})^{-1}R^*\bar{x}_t$$

if $\{\bar{x}_t\}$ is the only input. Show that if $\bar{x}_t$ is constant then this becomes $(I - \Gamma\mathcal{T})(x_t - \bar{x}) = 0$, as it must.

3.4 CONTROLLABILITY ETC.; INFINITE-HORIZON LIMITS

The concepts and results outlined in this section are important but standard.

We consider the question of horizon stability for the simple regulation problem of Section 3.2; conclusions for this imply conclusions for its variants. There are two basic lines of argument. One is that if stabilizing controls exist, then stationary policies with finite infinite-horizon cost also exist. The other is that if the infimal infinite-horizon cost is zero only for $x = 0$ (so that any deviation from $x = 0$ carries a cost) then the optimal policy is necessarily also stabilizing.

Suppose one started from an initial state x_0 and used the control $u_t = -Q^{-1}Sx_t$ which minimizes the actual control cost. The cost incurred over a time interval of length r would then be

$$\sum_0^{r-1} x_t'R^*x_t = x_0'\left(\sum_0^{r-1}(A^{*'})^t R^*(A^*)^t\right)x_0 \tag{3.32}$$

where $A^* = A - BQ^{-1}S$ and $R^* = R - S'Q^{-1}S$ are the values that A and R assume under the normalization to $S = 0$ (see Exercise 3.2.1). The use of any other control will imply an additional strictly positive control cost, so, if expression (3.32) is strictly positive for any non-zero x_0, then this implies that *any* initial deviation of state from zero incurs a positive cost over a time period of length r.

The condition that expression (3.32) should be positive for any non-zero x_0 for some r thus amounts to the requirement of *deviation sensitivity*: that any deviation of initial state from zero incurs a cost ultimately, whatever the policy. The condition implies that

$$\sum_{t=0}^{r-1} (A^{*\prime})^t R^* (A^*)^t > 0 \tag{3.33}$$

for some r (which either does not exist or is not greater than n. We shall express condition (3.33) by saying that R^* *is positive definite on* $\{(A^*)^t\}$.

An important condition on the plant equation (3.7) is that of *controllability*. This condition requires that for some r it should be possible to find a sequence of controls $u_0, u_1, \ldots, u_{r-1}$, taking the state variable from an arbitrary prescribed initial value $x_0 = d_1$ to an arbitrary prescribed terminal value $x_r = d_2$. Again, either such an r does not exist or it can be taken as not exceeding n. The controllability condition is equivalent to the requirement that the matrix

$$M_r = [B \quad AB \quad A^2 B \ldots A^{r-1} B] \tag{3.34}$$

should have rank n for some r, or, equivalently, that $BQ^{-1}B'$ should be positive-definite on $\{(A')^t\}$. Here Q is an $m \times m$ positive-definite matrix, its value otherwise immaterial. Note that these conditions are equivalent to those in which we substitute A^* for A.

A weaker condition than controllability is that of *stabilizability*: that there exists a matrix K such that $A + BK$ is a stability matrix. The point of this is that $A + BK$ is the gain matrix for the stationary policy $u_t = Kx_t$, and this policy will then be stabilizing if $A + BK$ is a stability matrix.

The conditions of deviation sensitivity and controllability are enough to ensure that the solution Π_t of the Riccati equation (3.13) converges to a limit Π as $t \to -\infty$ (or as $h \to +\infty$). More specifically, let us write $\Pi_t = \Pi_{(s)}$, where $s = h - t$ is time-to-go. We can then rewrite the Riccati equation (3.13) as

$$\Pi_{(s)} = f \Pi_{(s-1)} \qquad (s > 0) \tag{3.35}$$

where f is the operator with effect

$$f\Pi = R + A'\Pi A - (S' + A'\Pi B)(Q + B'\Pi B)^{-1}(S + B'\Pi A) \tag{3.36}$$

and $\Pi_{(0)}$ has the prescribed value Π_h.

Theorem 3.4.1. *Assume S normalized to zero. Assume the system sensitive to control deviation in that $Q > 0$, sensitive to state deviation in that R is positive definite on $\{A^t\}$, and controllable in that $J = BQ^{-1}B'$ is positive definite on $\{(A')^t\}$. Then:*

(i) *The equilibrium Riccati equation*

$$\Pi = f\Pi \tag{3.37}$$

has a unique non-negative-definite solution Π.

(ii) $\Pi_{(s)} \to \Pi$ *as s increases, for any finite and non-negative definite* $\Pi_{(0)}$.
(iii) *The limit gain matrix Γ is a stability matrix.*

For proof see e.g. Whittle (1982) Chapters 4 and 5. The matrix

$$J = BQ^{-1}B' \tag{3.38}$$

plays an important role. It represents, in a matrix sense, the ratio of control effectiveness in the plant equation (as measured by B) to control cost (as measured by Q). We shall term it the *control-power matrix*.

3.4.1 Exercises and comments

(1) Note an implication of the theorem: that controllability implies stabilizability. The form $u_t = Kx_t$ of control rule may seem too special to test stabilizability, when one considers that u could be chosen as any non-linear function of system history. However, the fact that the optimal infinite-horizon control (if one exists) is of this form justifies the choice.
(2) Note that the conditions of the theorem are by no means necessary. For example, suppose that $B = 0$ and A is a stability matrix. Then $\Pi = \lim \Pi_{(s)}$ exists and is the unique solution of $\Pi = R + A'\Pi A$, without condition on R and despite the fact that the system is not controllable. See also Section 3.10.

3.5 ALTERNATIVE FORMS OF THE RICCATI AND CONTROL RELATIONS

Theorem 3.5.1. *The expressions* (3.13), (3.14) *and* (3.15) *for the Riccati recursion and the optimal control matrix have the alternative forms*

$$\Pi_t = R^* + A^{*'}(J + \Pi_{t+1}^{-1})^{-1}A^* \tag{3.39}$$

$$K_t = -Q^{-1}S - Q^{-1}B'(J + \Pi_{t+1}^{-1})^{-1}A^* \tag{3.40}$$

Here J, A^* and R^* are just the control-power matrix and the normalized forms of A and R defined in the previous section. The alternative forms (3.39) and (3.40) often turn out to be useful. They correspond more immediately to the continuous-time forms we shall deduce in Section 3.9 than do the previous forms given in Section 3.2, and the spontaneous appearance of the control-power matrix is found to be meaningful. Equations (3.39) and (3.40) make sense even if Π_{t+1} is singular. In this case one would, for example, write

$$(J + \Pi^{-1})^{-1} = \Pi(I + J\Pi)^{-1}$$

Proof. We appeal to the useful identity

$$x'\Pi x = \max_{\lambda} (-2\lambda'x - \lambda'\Pi^{-1}\lambda) \tag{3.41}$$

which is the most explicit form of the Legendre (or maximum) transform. In terms of this we can write the Riccati recursion as

$$x'\Pi_t x = \underset{u,\lambda}{\text{stat}}\,[c(x,u) - 2\lambda'(Ax + Bu) - \lambda'\Pi_{t+1}^{-1}\lambda] \qquad (3.42)$$

Here $c(x,u)$ is the quadratic cost function (3.9) and 'stat' indicates 'stationary value'—in this case a maximum with respect to λ and a minimum with respect to u.

If we maximize out λ in equation (3.42) we, of course, reduce the recursion to its original form (3.16); the minimization with respect to u then yields the forms for the Riccati recursion and the optimal control rule stated in Theorem 3.2.1. However, if we invert the order of taking these extremes then we deduce the alternative forms (3.39) and (3.40), as we leave the reader to verify. This inversion is valid, because the bracketed expression in (3.42) has a unique stationary point: a saddle point. □

The introduction of the supplementary variable λ is a half-way step towards a formulation of the maximum principle, a step we now complete.

3.5.1 Exercise and comments

(1) Note the expression $\Gamma_t = (I + J\Pi_{t+1})^{-1}A^*$.

3.6 APPEAL TO THE MAXIMUM PRINCIPLE

For the deterministic model of Section 3.2 one can regard the control-optimization problem as the minimization of the cost function defined by equations (3.8)–(3.10) subject to the constraint of the plant equation (3.7), this being regarded as a vector constraint on the path $\{x_t, u_t\}$ at every value of t. One could take account of this constraint by the introduction of Lagrangian multipliers, and so consider the minimization of the path integral

$$\mathbb{I}(x,u,\lambda) = \sum_{t=0}^{h-1} c(x_t, u_t) + \mathbb{C}_h(x_h) + 2\sum_{t=1}^{h} \lambda'_t(x_t - Ax_{t-1} - Bu_{t-1}) \qquad (3.43)$$

Here c and $\mathbb{C}_h$ have the quadratic forms assumed in (3.9) and (3.10), and λ_t is an n-vector, the multiplier corresponding to the constraint

$$x_t = Ax_{t-1} + Bu_{t-1} \qquad (0 < t \leq h) \qquad (3.44)$$

Let us formalize the statements that can be made.

Theorem 3.6.1. *The maximum principle: Assume the control-optimization problem well posed, in that the cost function satisfies the positivity conditions of* (M3) *(Section 3.2). Then values of the multipliers (conjugate variables) λ_t exist such that* the solution of the control-optimization problem over $[0, h]$ can be obtained by the minimization of $\mathbb{I}(x, u, \lambda)$ with respect to $\{x_t, u_t\}$, subject only to the prescription of x_0. The appropriate values of the multipliers λ are those which maximize $\mathbb{I}(x, u, \lambda)$; the optimal solution $\{x_t, u_t, \lambda_t\}$ is thus the unique saddle point of $\mathbb{I}$.

The assertions can be regarded as an appeal to the Pontryagin maximum principle, which can be formally derived by dynamic programming arguments (see e.g. Whittle, 1982, p. 104). More simply, we can just regard them as an appeal to the theory of convex programming, justified by the convexity of the cost function $\mathbb{C}$, the linearity of the constraints (3.44) and the convexity of the set within which $\{x_t, u_t\}$ may be chosen. The values of the multipliers λ can be regarded as being fixed either by constraints (3.44) or, equivalently, by the 'dual' characterization that the path integral $\mathbb{I}$ be minimal with respect to the x and u variables and maximal with respect to the λ variables at the optimal solution.

This min–max (saddle point) can be located stationarity conditions in the present case. That with respect to λ_t yields just the plant equation (3.44), and those with respect to x_t and u_t yield

$$Rx_t + S'u_t + \lambda_t - A'\lambda_{t+1} = 0 \qquad (0 < t < h) \tag{3.45}$$

$$Sx_t + Qu_t - B'\lambda_{t+1} = 0 \qquad (0 \leq t < h) \tag{3.46}$$

These equations are to be supplemented with the terminal condition

$$\Pi_h x_h + \lambda_h = 0 \tag{3.47}$$

and the initial condition: prescription of x_0.

Relations (3.44)–(3.46) determine (x, u, λ) along an optimal trajectory; they have a significance which we shall begin to develop in the next section. The relation between this approach and that of recursive dynamic programming can be seen by calculating a partial extreme. We have

$$\underset{x_{t+2}\ldots, u_{t+1}\ldots, \lambda_{t+2}\ldots}{\text{stat}} \mathbb{I} = \sum_{j=0}^{t} c(x_j, u_j) + \sum_{j=1}^{t+1} \lambda'_j(x_j - Ax_{j-1} - Bu_{j-1}) + x'_{t+1}\Pi_{t+1}x_{t+1} \tag{3.48}$$

where the 'stat' operator implies that the variables indicated have been extremized out of $\mathbb{I}$. We thus have

$$x'_t \Pi_t x_t = \underset{x_{t+1}, u_t, \lambda_{t+1}}{\text{stat}} [c(x_t, u_t) + 2\lambda'_{t+1}(x_{t+1} - Ax_t - Bu_t) + x'_{t+1}\Pi_{t+1}x_{t+1}] \tag{3.49}$$

The x_{t+1}-minimization yields the condition $\Pi_{t+1}x_{t+1} + \lambda_{t+1} = 0$ and reduces (3.49) to relation (3.42), from which we know that both forms of the Riccati recursion and the optimal control can be derived. We are thus back on familiar territory, and have established, furthermore, that relation (3.47) holds for all t on an optimal orbit:

$$\Pi_t x_t + \lambda_t = 0 \qquad (0 < t \leq h) \tag{3.50}$$

3.7 THE HAMILTONIAN FORMULATION

In the previous section we considered optimization over the time interval $[0, h]$. Let us now take the generic situation: that one has reached time t, not necessarily by an optimal path, and wishes to optimize remaining decisions $u_t, u_{t+1}, \ldots$. Let us define the vector

$$\zeta = \begin{bmatrix} x \\ u \\ \lambda \end{bmatrix} \tag{3.51}$$

and denote the value of ζ at time τ under optimal operation from time t by $\zeta_\tau^{(t)}$. Then equations (3.44)–(3.46) could be written

$$\Phi(\mathcal{T})\zeta_\tau^{(t)} = 0 \qquad (t \leq \tau < h) \tag{3.52}$$

where $\mathcal{T}$ operates on the *subscript* τ and $\Phi(\mathcal{T})$ is the matrix of operators

$$\Phi(\mathcal{T}) = \begin{bmatrix} R & S' & I - A'\mathcal{T}^{-1} \\ S & Q & -B'\mathcal{T}^{-1} \\ I - A\mathcal{T} & -B\mathcal{T} & 0 \end{bmatrix} \tag{3.53}$$

In anticipation of the more general cases to come we shall write (3.53) as

$$\Phi(\mathcal{T}) = \begin{bmatrix} R & S' & \bar{\mathcal{A}} \\ S & Q & \bar{\mathcal{B}} \\ \mathcal{A} & \mathcal{B} & 0 \end{bmatrix} \tag{3.54}$$

Here $\mathcal{A}$ and $\mathcal{B}$ are the operators occurring in the general plant equation

$$\mathcal{A}x + \mathcal{B}u = 0 \tag{3.55}$$

If, for example,

$$\mathcal{A} = \sum_j A_j \mathcal{T}^j \tag{3.56}$$

then we define the adjoint operator $\bar{\mathcal{A}}$ by

$$\bar{\mathcal{A}} = \sum_j A'_j \mathcal{T}^{-j} \tag{3.57}$$

Equations (3.52) (plus boundary conditions, which we shall discuss presently) determine the optimal course of the process from an arbitrary history before time t. The operator matrix has a pleasing form, Hermitian in that

$$\bar{\Phi} = \Phi \tag{3.58}$$

This is the property we refer to as *Hamiltonian*, for reasons explained in Appendix 4. The Hamiltonian equations (3.52) follow from the fact that we have been able to characterize the optimal path as one that renders a quadratic path integral stationary. This is a characterization that persists under powerful generalizations of the problem, and constitutes the programme of Part III. It is also an approach which provides powerful methods for complete solution.

The system (3.52) is subject to the initial conditions: prescription of x_{t-1} and u_{t-1}. (The variable $\lambda_\tau^{(t)}$ does not occur for $\tau < t$.) The terminal conditions can be expressed

$$\Pi_h x_h^{(t)} + \lambda_h^{(t)} = 0$$
$$x_h^{(t)} = A x_{h-1}^{(t)} + B u_{h-1}^{(t)} \tag{3.59}$$

There would also be an equation for $u_h^{(t)}$ if u_h appeared in the terminal cost. Therefore the two first equations of (3.52) (implying stationarity of the path integral with respect to x and u) are now modified, because of the change in the form of the instantaneous cost function at termination and because, formally,

$$\lambda_\tau^{(t)} = 0 \qquad (\tau > h) \tag{3.60}$$

The final relation of (3.52), the plant equation, persists.

The Hamiltonian approach is particularly powerful in the infinite-horizon limit, when equations (3.52) will hold for $\tau \geq t$. If the problem is horizon stable then the optimal control will be stationary and $\zeta_\tau^{(t)}$ will converge to zero exponentially fast with increasing τ. This convergence replaces the terminal conditions (3.59) and (3.60).

Consider the matrix-generating function $\Phi(z)$, where z is a complex scalar. Assume that this can be factorized:

$$\Phi(z) = \Phi_-(z)\Phi_0\Phi_+(z) \tag{3.61}$$

where Φ_0 is a constant matrix (i.e. independent of z), Φ_+ is such that both $\Phi_+(z)$ and $\Phi_+(z)^{-1}$ have an expansion in non-negative powers of z which is valid within and on the unit circle (i.e. for $|z| \leq 1$), and Φ_- is such that both $\Phi_-(z)$ and $\Phi_-(z)^{-1}$ have an expansion in non-positive powers of z which is valid outside and on the unit circle. The Hermitian character of Φ will in fact imply that the factorization can be so chosen that Φ_0 is symmetric and

$$\Phi_- = \bar{\Phi}_+ \tag{3.62}$$

The constant factor Φ_0 could have been absorbed in the other factors, but we shall have reason to leave it distinct.

With equation (3.52) holding for $\tau \geq t$ and $\zeta_\tau^{(t)}$ tending exponentially fast to zero with increasing τ we can properly divide out the factor $\Phi_-\Phi_0$ of Φ (see Exercise 3.7.2 (2) for details) and be left with the equation

$$\Phi_+(\mathcal{T})\zeta_\tau^{(t)} = 0 \qquad (\tau \geq t) \tag{3.63}$$

3.7 THE HAMILTONIAN FORMULATION

This equation gives the stable forward dynamic recursion for the optimally controlled process, and determination of the factor $\Phi_+(z)$ in fact determines the optimal infinite-horizon control.

We shall see later (Section 11.1) that, in the Markov case we are now considering, the canonical factorization (3.61) can be achieved with

$$\Phi_+(\mathcal{T}) = \begin{bmatrix} \Pi & 0 & I \\ -K & I & 0 \\ \mathcal{A} & \mathcal{B} & 0 \end{bmatrix} \tag{3.64}$$

$$\Phi_0 = \begin{bmatrix} 0 & 0 & I \\ 0 & \tilde{Q} & 0 \\ I & 0 & -\Pi \end{bmatrix} \tag{3.65}$$

where Π and K are the infinite-horizon limits of the value function and control matrices of Theorem 3.2.1, and $\tilde{Q} = Q + B'\Pi B$. Determination of Φ_+ thus indeed determines both Π and K.

If we take equation (3.63) at $\tau = t$ with Φ_+ having the determination (3.64) then the three relations expressed are the λ/x relationship

$$\Pi x_t + \lambda_t = 0$$

(c.f. (3.50)), the optimal control rule

$$u_t = K x_t$$

and the plant equation itself.

Consider now the non-homogeneous regulation problem of Section 3.3, with the plant equation taking the disturbed form (3.23) and the cost function the variable set-point form (3.28). It is then a direct matter to verify that equation (3.52) and the first equation of (3.59) become modified to

$$\Phi(\mathcal{T})\zeta_\tau^{(t)} = \omega_\tau \qquad (t \leq \tau < h) \tag{3.66}$$

$$\Pi_h(x_h^{(t)} - \bar{x}_h) + \lambda_h^{(t)} = 0 \tag{3.67}$$

where ω_τ is the known vector

$$\omega_\tau = \begin{bmatrix} R\bar{x}_\tau + S'\bar{u}_\tau \\ S\bar{x}_\tau + Q\bar{u}_\tau \\ \alpha_\tau \end{bmatrix} \tag{3.68}$$

Then, by the same arguments as led us from (3.52) to (3.63), assuming appropriate limit behaviour for infinite τ, one can properly perform the partial inversion of relation (3.66), if h is infinite, to

$$\Phi_+(\mathcal{T})\zeta_\tau^{(t)} = \Phi_0^{-1}\Phi_-(\mathcal{T})^{-1}\omega_\tau \qquad (\tau \geq t) \tag{3.69}$$

Equation (3.69) at $\tau = t$ yields the optimal control at time t, explicitly and in closed-loop form. Moreover, the feedback/feedforward form of the rule is clear. In equation (3.69) the factor $\Phi_+(\mathcal{T})$ is to be expanded in non-negative powers of $\mathcal{T}$, and so operates into the past. The factor $\Phi_-(\mathcal{T})^{-1}$ is to be expanded in non-positive powers of $\mathcal{T}$, and so operates into the future. The operator $\Phi_+(\mathcal{T})$ supplies feedback; the operator $\Phi_-(\mathcal{T})^{-1}$ supplies feedforward.

We now see the point of the observation made in Section 3.3: the dynamics of feedback and feedforward are mutually adjoint in that they are generated, respectively, by the two mutually adjoint factors Φ_+ and Φ_- of Φ. We follow up the calculations in the exercises to show that relation (3.69) does indeed imply the particular relation (3.27), for example.

The conditions on controllability, etc. that we required for Theorem 3.4.1 are sufficient to ensure existence of the canonical factorisation (3.61). That this factorization should exist constitutes exactly the necessary and sufficient condition that the regulation process of Section 3.1 should be horizon stable. For the non-homogeneous case of Section 3.3 we will require also that the sequence $\{\omega_t\}$ should be well enough behaved that the linear function of future ω-values $\Phi_-(\mathcal{T})^{-1}\omega_t$ is convergent.

We shall return to factorisation (3.61) and its natural forms when we consider the more general models of Part III. This will also provide the occasion to investigate the relation between the canonical factorization and the solution of the Riccati equation.

3.7.1 Exercises and comments

(1) Verify that the matrices (3.64) and (3.65) achieve the canonical factorization (3.61). Note that matrix (3.64) can be factored to

$$\begin{bmatrix} I & 0 & 0 \\ 0 & I & 0 \\ 0 & -Bz & I \end{bmatrix} \begin{bmatrix} \Pi & 0 & I \\ -K & I & 0 \\ I - \Gamma z & 0 & 0 \end{bmatrix}$$

and hence that its canonical character is equivalent to the statement that $\Gamma = A + BK$ is a stability matrix.

(2) *Finite-horizon results.* We can write the canonical factorization (3.64) rather in the form

$$\Phi(z) = (I - \gamma' z^{-1})\beta(I - \gamma z)$$

where γ is a stability matrix and β symmetric. Consider now equation (3.52); let us, for simplicity, dispense with the t superscript. By solving first for $\beta(I - \gamma\mathcal{T})\zeta_\tau$ and then for ζ_τ we find that this has the general solution

$$\zeta_\tau = \gamma^{\tau-t+1}\zeta_{t-1} + D_\tau(\gamma')^{h-\tau}\delta \qquad (3.70)$$

where

$$D_\tau = \sum_{j=0}^{\tau-t} \gamma^j \beta^{-1}(\gamma')^{-j}$$

3.7 THE HAMILTONIAN FORMULATION

and δ is a vector to be determined from terminal conditions. The terminal conditions amount in the present (homogenous) case to equation (3.59). Let us write these as

$$M_1 \zeta_h + M_2 \zeta_{h-1} = 0$$

so determining δ as

$$\delta = -(M_1 D_h + M_2 D_{h-1} \gamma')^{-1} (M_1 \gamma + M_2) \gamma^{h-t} \zeta_{t-1} \qquad (3.71)$$

With this the solution (3.70) of (3.52) for the given end-conditions is fully determined. It is evident from equation (3.70) and (3.71) that we have in fact

$$\zeta_\tau = \gamma^{\tau - t + 1} \zeta_{t-1} + O(\rho^{2h - t - \tau}) \qquad (t \leq \tau < h)$$

where ρ is the modulus of the maximal-modulus eigenvalue of γ. The fact that the final term in this solution tends to zero with increasing h legitimizes the replacement of (3.52) by (3.63), i.e. by

$$(I - \gamma \mathcal{T}) \zeta_\tau = 0 \qquad (\tau \geq t)$$

in the infinite-horizon case.

(3) Consider the disturbed regulation problem of Section 3.3 in the case $\bar{x} = 0$, $\bar{u} = 0$, to which we know we can normalize. Demonstrate, then, that, for the particular factorization (3.64)/(3.65), relation (3.69) becomes, at $\tau = t$,

$$\Phi_+(\mathcal{T}) \zeta_t = \begin{bmatrix} (I - \bar{\mathcal{G}}^{-1}) \Pi \alpha_t \\ -\tilde{Q}^{-1} B' \bar{\mathcal{G}}^{-1} \Pi \alpha_{t+1} \\ \alpha_t \end{bmatrix}$$

where $\mathcal{G} = (I - \Gamma \mathcal{T})$. The second row of this equation gives exactly the feedback/feedforward determination (3.27) of the optimal control law. The third row gives just the plant equation. The first gives

$$\Pi x_t + \lambda_t = \Pi \alpha_t - \sum_{j=0}^{\infty} (\Gamma')^j \Pi \alpha_{t+j}$$

corresponding to

$$\Pi x_t - \sigma_t + \lambda_t = 0$$

(4) The equation system (3.52) can be reduced by elimination. If $Q > 0$ then we can eliminate u to obtain

$$\Phi_1(\mathcal{T}) \begin{bmatrix} x \\ \lambda \end{bmatrix}_\tau^{(t)} = 0 \qquad (3.72)$$

where

$$\Phi_1(\mathcal{T}) = \begin{bmatrix} R^* & I - A^{*'} \mathcal{T}^{-1} \\ I - A^* \mathcal{T} & -J \end{bmatrix}$$

Note the spontaneous appearance (again) of the control power matrix J and the normalized versions of A and R defined in Section 3.4. The factorizability of the reduced matrix operator Φ_1 expresses exactly the weakened versions of the controllability and deviation-sensitivity conditions of Theorem 3.4.1 that are necessary and sufficient for the conclusions of that theorem.

(5) Let us assume S normalized to zero, so that we can write A^* simply as A, etc. If we are willing to assume that $R > 0$ as well as $Q > 0$ then we can also eliminate x from system (3.72) and obtain the further-reduced system

$$\Phi_2(\mathcal{T})\lambda_\tau^{(t)} = 0$$

where

$$\Phi_2(z) = [(I - Az)R^{-1}(I - A'z^{-1}) + BQ^{-1}B']$$

Factorizability of Φ_2 must be closely equivalent to controllability of the system, and, indeed, is exactly equivalent. $\Phi_2(z)$ will not have a canonical factorization if $|\Phi_2(z)|$ has a zero on the unit circle. If it has such a zero then

$$\eta'\Phi_2(z) = 0$$

for some z on the unit circle and non-zero vector η. Since R^{-1} and Q^{-1} are positive-definite and z^{-1} is the conjugate of z this implies that

$$\eta'B = 0, \qquad \eta' = z\eta'A$$

However, these relations imply that $\eta'M_r = 0$ for any r, where M_r is the matrix defined in (3.34), and so imply the system to be uncontrollable. Conversely, controllability implies that $|\Phi_2(z)|$ cannot have a zero on the unit circle.

3.8 NUMERICAL METHODS

The discussion hitherto has reduced the determination of the infinite-horizon optimal control to one of two basic (and presumably equivalent) problems. These are: the solution of the equilibrium Riccati equation

$$\Pi = f\Pi \tag{3.73}$$

for Π, where the operator f is defined in (3.36), or the canonical factorization of the matrix of polynomials $\Phi(z)$ defined in (3.53). In this section we consider only the former problem, although the method we ultimately fix on also provides an effective and natural solution for the latter (see Section 12.4).

We know that, under the conditions of Theorem 3.4.1, the relevant solution of (3.73) (the unique non-negative definite solution) is also the limit for large s of the sequence of matrices $\Pi_{(s)}$ generated by the recursion

$$\Pi_{(s)} = f\Pi_{(s-1)} \qquad (s > 0) \tag{3.74}$$

starting from a non-negative definite $\Pi_{(0)}$. Calculation of this sequence corre-

3.8 NUMERICAL METHODS

sponds to the method of *value iteration*: the determination of optimal policy over successively increasing horizons. As a method of calculating the limit value Π it can be slow to converge. Under unfavourable conditions rounding errors can even lead to a loss of positive definiteness. For this reason, the so-called *square-root* method is often recommended; for an account see e.g. Morf and Kailath (1975), Silverman (1976) or Whittle (1982).

A method which gives much faster convergence than either value iteration or the square root method is that of *policy improvement*. For notational simplicity we shall use K_i, Γ_i and Π_i to denote the values of K, Γ and Π determined at the ith stage of policy improvement. These are not to be confused with K_t, Γ_t and Π_t; we invariably use i as the subscript denoting stage of iteration.

For policy improvement generally see the references in Section 3.11. For the LQ case the actual calculations are as follows. At the ith stage of iteration one has a stationary policy

$$u_t = K_i x_t \qquad (3.75)$$

To assume a control of this form (linear and Markov) is no restriction, since the optimal infinite-horizon control is known to be of this form.

Suppose policy (3.75) stabilizing, so that the matrix $\Gamma_i = A + BK_i$ is a stability matrix. The infinite-horizon value function associated with this policy is then finite and equal to $x'\Pi_i x$, where Π_i satisfies

$$\Pi_i - \Gamma_i' \Pi_i \Gamma_i = c(K_i) \qquad (3.76)$$

and

$$c(K_i) = R + K_i' S + S' K_i + K_i' Q K_i$$

(c.f. Exercise 3.2.4). One then determines an improved control by determining K_{i+1} as the value of K minimizing $c(K) + (A + BK)' \Pi_i (A + BK)$, namely

$$K_{i+1} = -(Q + B'\Pi_i B)^{-1}(S + B'\Pi_i A) \qquad (3.77)$$

The policy $u_t = K_{i+1} x_t$ will then be an improved policy in that $\Pi_{i+1} \leq \Pi_i$, with equality iff policy (3.75) is optimal.

This assertion implies that if one can find a stabilizing policy with which to start the algorithm, then the sequence $\{\Pi_i\}$ decreases with increasing i to the solution Π of (3.73) associated with the infinite-horizon optimal policy. One expects that the sequence $\{\Pi_i\}$ will converge to Π faster than does the sequence $\{\Pi_{(s)}\}$ generated by the value iteration algorithm, because the policy improvement algorithm evaluates an infinite-horizon policy at every stage. However, the contrast can be expressed even more convincingly.

Theorem 3.8.1. *The method of value iteration corresponds to the solution of equation (3.73) by the iteration (3.74); the method of policy improvement corresponds to its solution by the Newton–Raphson algorithm.*

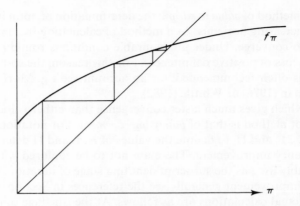

Figure 1. Convergence to the solution of $\Pi = f\Pi$ by value iteration

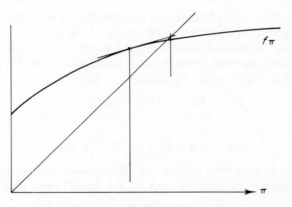

Figure 2. Convergence to the solution of $\Pi = f\Pi$ by policy improvement

It is the second assertion which is the substantial one, and makes plain why the policy-improvement algorithm should converge so much more rapidly. We illustrate the one-dimensional case graphically in Figure 1. The matrix function $f\Pi$ has the increasing and concave character indicated, and the positive crossing point of the graph with the straight-line graph of Π itself locates the root of (3.73).

Value iteration (the iteration of f) corresponds to the staircase construction of Figure 1. Policy improvement (i.e. the use of the Newton–Raphson algorithm) replaces the staircase construction of Figure 1 by passage along a tangent to $f\Pi$ (see Figure 2) and is evidently much faster. However, for convergence it is necessary that one starts from a value Π_0 at which $f\Pi$ has derivative less than

3.8 NUMERICAL METHODS

unity. This is related to the point that policy improvement must start from a stabilizing policy.

Proof. The first assertion of the theorem merely repeats the definition of value iteration; it is the second assertion we have to prove. The Newton–Raphson method of improving the solution of (3.73) from a trial value Π_i is to substitute $\Pi_i + \Delta$ for Π in (3.73) and neglect all powers of Δ higher than the first. The equation would then take the form

$$\Pi_i + \Delta = f\Pi_i + L(\Pi_i, \Delta) \tag{3.78}$$

where $L(\Pi_i, \Delta)$ is the linear functional of Δ that one obtains from the first-order term in the expansion of $f(\Pi_i + \Delta)$; its determination implies a determination of the form of the matrix derivative of $f\Pi$ at Π_i. One then sets $\Delta = \Pi_{i+1} - \Pi_i$, and so obtains a linear equation for Π_{i+1}.

It follows from the extremal representation of f

$$f\Pi = \inf_K (c(K) + \Gamma'\Pi\Gamma) \tag{3.79}$$

(with $\Gamma = A + BK$) that

$$f\Pi_i = c(K_{i+1}) + \Gamma'_{i+1}\Pi_i\Gamma_{i+1}$$

$$L(\Pi_i, \Delta) = \Gamma'_{i+1}\Delta\Gamma_{i+1}$$

where K_{i+1} is, as in (3.77), the minimizing value in (3.79). Substituting these expressions into (3.78) and setting $\Delta = \Pi_{i+1} - \Pi_i$ we derive the equation

$$\Pi_{i+1} - \Gamma'_{i+1}\Pi_{i+1}\Gamma_{i+1} = c(K_{i+1})$$

determining Π_{i+1} in terms of K_{i+1}. However, this is just equation (3.76) written at $i+1$ instead of i, and we have already seen that K_{i+1} is determined in terms of Π_i by the prescription (3.77). The equivalence of the policy-improvement and Newton–Raphson algorithms is thus established. □

It would follow from general theorems on the Newton–Raphson method that, if the method converges, it shows second-order convergence. (Roughly, convergence of order r is equivalent to $\Delta_{i+1} = O(\Delta_i^r)$.) The equivalence established in Theorem 3.8.1 implies that, if one starts from a value Π_0 which corresponds to the value function of a stabilizing policy, then convergence indeed occurs and is monotonic. The equivalence also gives the Newton–Raphson method a natural variational significance in this context. The following theorem on rates of convergence can be established directly.

Theorem 3.8.2. *The methods of value iteration and policy improvement show, respectively, first- and second-order convergence to the limit matrix* Π.

To form an idea of the actual speeds of the two algorithms consider the case $n = 2, m = 1$, with

$$A = \begin{bmatrix} 1 & 1 \\ 0 & 1 \end{bmatrix} \quad B = \begin{bmatrix} 0 \\ 1 \end{bmatrix}$$

$$R = \begin{bmatrix} 1 & 0 \\ 0 & 4 \end{bmatrix} \quad Q = 1 \quad S = 0$$

The optimal values of Π and K for this case turn out to be

$$\Pi = \begin{bmatrix} 3.715193 & 3.043738 \\ 3.043738 & 8.264339 \end{bmatrix}$$

$$K = [-0.328543 \quad -1.220601]$$

to seven significant figures. Starting from a trial value $\Pi_{(0)} = 0$ the value iteration method required 12 steps to calculate Π to four-figure accuracy, 19 steps for seven-figure accuracy. In applying the policy-improvement method one cannot simply set $\Pi_0 = 0$; one must rather choose an initial value of the control matrix K which ensures stability. The value chosen was $[-0.25 \quad -1.00]$, a value chosen before the optimal value given above had been determined. With this starting value the policy-improvement method evaluated Π to four-figure accuracy in two steps, to seven-figure accuracy in four steps. The evidence for greatly increased speed is thus convincing, despite the fact that the starting points of the two methods are not strictly comparable.

The assertions of Theorems 3.8.1 and 3.8.2 are important to us in the sequel because they extend to the higher-order and risk-sensitive cases. The Riccati recursion survives in some form under both these generalizations, and the theorems continue to apply. However, the Hamiltonian formulation of the previous section also extends, with the problem of solving the equilibrium Riccati equation translated into the more elegant and useful form: the canonical factorization of an operator. The Newton–Raphson method also translates to provide a beautiful iterative method of achieving this factorization, a method which, Theorems 3.8.1 and 3.8.2 tell us, has a variational basis and shows second-order convergence.

It is natural to ask whether there are any other optimization models for which the optimality equation reduces to a parameter recursion (e.g. an analogue of the Riccati equation) and for which policy improvement is equivalent to application of the Newton–Raphson algorithm to this recursion. The point has been investigated (Whittle, 1989a). A larger class of such models is found, of which, however, only the LQG and LEQG models are interesting.

3.9 CONTINUOUS TIME

The continuous-time version of an LQ model for regulation to zero with a deterministically disturbed plant would have plant equation

$$\dot{x} = Ax + Bu + \alpha \tag{3.80}$$

3.9 CONTINUOUS TIME

and cost function

$$\mathbb{C} = \int_0^\infty c(x,u)\,dt + \mathbb{C}_h(x(h)) \tag{3.81}$$

where $c(x, u)$ and $\mathbb{C}_h(x)$ have again the forms (3.9) and (3.10).

One can obtain the continuous-time results as a formal limit of those for discrete time by letting the time increment between consecutive observation/action points be δ rather than unity, replacing the quantities A, B, R, S, Q and α of the discrete-time treatment by $I + A\delta$, $B\delta$, $R\delta$, $S\delta$, $Q\delta$ and $\alpha\delta$, respectively, and then letting δ tend to zero.

We find then, as in (3.24), that the value function has the quadratic form in x:

$$F(x, t) = x'\Pi x - 2\sigma'x + \gamma \tag{3.82}$$

where Π, σ and γ are functions of t. The two forms (3.13)/(3.14) and (3.39) of the Riccati equation for Π have the same continuous analogue:

$$\dot{\Pi} + R + A'\Pi + \Pi A - (S' + \Pi B)Q^{-1}(S + B'\Pi) = 0 \tag{3.83}$$

This can also be written

$$\dot{\Pi} + R^* + A^{*\prime}\Pi + \Pi A^* - \Pi J \Pi = 0 \tag{3.84}$$

where, as in discrete time, J is the control-power matrix $BQ^{-1}B'$ and $A^* = A - BQ^{-1}S$ and $R^* = R - S'Q^{-1}S$ are the transformed forms of A and R under the normalization $S = 0$.

The continuous-time versions of the Riccati equation are simpler, in that $(Q + B'\Pi B)^{-1}$ is replaced simply by Q^{-1}. One can say that in a short time interval of length δ the control chosen can affect the gradient of the future value function by an amount of order δ, but the curvature only by an amount of order δ^2. It is for this reason also that both discrete-time forms of the Riccati equation coalesce.

The optimal control is

$$u = Kx + Q^{-1}B'\sigma \tag{3.85}$$

(c.f. (3.26)) where the time-dependent matrix K is given by

$$K = -Q^{-1}(S + B'\Pi) \tag{3.86}$$

(c.f. (3.15)) and σ obeys the equation

$$\dot{\sigma} + \Gamma'\sigma - \Pi\alpha = 0 \tag{3.87}$$

(c.f. (3.25)) plus a prescription of terminal σ. Here $\Gamma = A + BK$, as ever. We expect Γ to converge with increasing horizon to a stability matrix, which means, in the continuous-time context, that $\exp(t\Gamma)$ converges to zero with increasing t, or that all eigenvalues of Γ have a strictly negative real part.

The criteria for controllability and deviation sensitivity have expressions analogous to those of the discrete-time case. For example, the criterion for controllability over a time interval of length t is that

$$\int_0^t e^{A\tau} B Q^{-1} B' e^{A'\tau} \, d\tau > 0 \tag{3.88}$$

and validity of (3.88) for some $t > 0$ implies validity for all $t > 0$. However, the smaller the value of t allowed, the larger will be the values of u required to bring about the required transition (see Exercise 3.9.1).

The material of Sections 3.7 and 3.8, on the Hamiltonian approach and on numerical methods, also has an analogue, which we shall pursue in Chapters 15 and 16.

3.9.1 Exercise and comments

(1.) Suppose that one wishes to choose u to take the state variable from prescribed $x(0)$ to prescribed $x(t)$ in such a way as to minimize $\int_0^t u'Qu \, d\tau$. Show that this is achieved by a control

$$u(\tau) = Q^{-1} B' e^{A'(t-\tau)} G(t)^{-1} [x(t) - e^{At} x(0)] \qquad (0 \le \tau < t)$$

where $G(t)$ is the matrix in the left-hand member of (3.88).

3.10 VIOLATIONS OF THE POSITIVE-DEFINITENESS CONDITIONS

One can easily find practical problems for which the positive-definiteness conditions of condition M3 (Section 3.2) are violated. These can still be well posed and even show horizon stability, although the results no longer have quite the simplicity of Theorem 3.4.1.

As an example, consider a scalar problem in which x represents the market price of a commodity and u the amount of that commodity manufactured by a monopolist. A plausible continuous-time plant equation is

$$\dot{x} = a(p - x - u) \tag{3.89}$$

This implies that if production were held at a constant value u, then the price would adjust itself to an equilibrium value $x = p - u$ at a proportional rate a. The monopolist receives a profit per unit time of $xu - cu - \frac{1}{2}u^2$, the three terms in this expression representing, respectively, receipts from sale, production costs and an increase in unit production costs with production level. To be consistent with our formulation hitherto, we shall work in terms of costs rather than of gains, and so shall set

$$c(x, u) = u^2 + 2cu - 2xu \tag{3.90}$$

(Multiplication by a factor of 2 is convenient.) The problem is not completely LQ, in that neither x nor u can become negative (or, at least, if they become so, then radically different mechanisms would prevail). However, we shall neglect

3.10 VIOLATIONS OF THE POSITIVE-DEFINITENESS CONDITIONS

this point, essentially assuming that we shall be operating in a region of state space for which both x and u remain positive under optimal rules.

The problem is then non-homogeneous LQ and the value function will have the quadratic form (3.82). It follows routinely (c.f. (3.83)) that the equation for Π is

$$\mathring{\Pi} = -1 - 4a\Pi - a^2\Pi^2 \tag{3.91}$$

where $\mathring{\Pi} = d\Pi/ds = -d\Pi/dt$; the rate of change of Π with time-to-go, $s = h - t$. For the moment we shall consider only this equation for Π; in Section 7.8 we shall give a complete analysis for the risk-sensitive case in which we consider also the form of the optimal control rule.

Equation (3.91) has two equilibrium solutions, both of them *negative*:

$$\Pi = a^{-1}(-2 \pm \sqrt{3}) \tag{3.92}$$

If we write equation (3.91) as $\mathring{\Pi} = \chi(\Pi)$ then the graph of $\chi(\Pi)$ is as in Figure 3. The stable Π-solution (i.e. that to which Π will converge with increasing s) is that for which $\chi'(\Pi)$ is negative. This is the upper solution $\Pi^{(2)} = a^{-1}(-2 + \sqrt{3})$. Even this is the limit value only if the zero horizon value $\Pi_{(0)}$ of Π exceeds the other root, $\Pi^{(1)} = a^{-1}(-2 - \sqrt{3})$. If $\Pi_{(0)} < \Pi^{(1)}$ then the value of Π decreases indefinitely with increasing s.

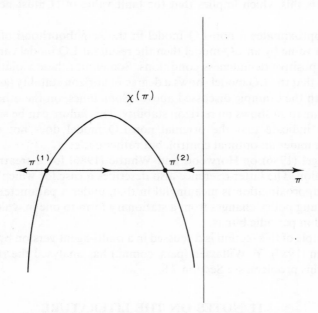

Figure 3. *Solution of the equilibrium Riccati equation for the production problem. The solutions for Π are negative, the larger value being the stable one*

Therefore the limit value of Π (in cases where one exists) does not have the positive value one would have expected under the standard assumptions of Theorem 3.4.1, but is negative. This is a consquence of the mixed character of the quadratic form (3.90), which is positive definite in u for fixed x (so assuring existence of a finite optimizing control) but decreases indefinitely as x increases for fixed positive u.

We should really be thinking in terms of gains rather than of losses. The fact that $-c(x, u)$ and $-F = -\Pi x^2 + 2\sigma' x + \gamma$ both increase with x for x large enough just reflect the fact that increasing prices are beneficial for the monopolist. The fact that $-F$ also increases as x *decreases* to $-\infty$ must be regarded as an artefact; the consequence of moving into regions of negative x and u for which the model is no longer realistic. Similarly, the fact that $-\Pi$ (and so $-F$) diverges to $+\infty$ with increasing s if the terminal value $-\Pi_{(0)}$ is sufficiently large again reflects behaviour under circumstances for which the model is unrealistic. The literal interpretation is that, if the value attached to a high terminal price is high enough, then the monopolist will produce at a negative rate (i.e. purchase his product back from the market) and so drive price up. The loss incurred by doing this is, if $-\Pi_{(0)} > -\Pi^{(1)}$, more than balanced by the increase in terminal reward.

Thus some features of the model are grossly unrealistic. However, the increasing character of $-F$ with x as x become sufficiently large and positive is not, and it is this which implies that the limit value of Π must necessarily be negative.

If one approximates a non-LQ model in the neighbourhood of its optimal equilibrium value by an LQ model then the resultant LQ model can very easily violate the positive-definiteness conditions. Sometimes these violations are acceptable, in that the LQ model shows a degree of horizon stability (as was indeed the case with the example discussed above). Sometimes, on the other hand, the LQ approximation shows no horizon stability. This failure can be significant, in that it can indicate that the original non-LQ model does not settle to an equilibrium under an optimal control, but rather cycles.

See Stengel (1986) or Horwood and Whittle (1986) for a treatment of LQ approximation. The latter reference also describes a case for which breakdown of the LQ approximation is meaningful in that, under a parameter change, an optimal fishing policy changes from a stationary form to one in which the stock is harvested in periodic bursts.

The example of this section is discussed in a multi-agent version by Fershtman and Kamien (1987). Y. Willassen (pers. comm.) has analysed the risk-sensitive version of this problem; see Section 7.8.

3.11 NOTES ON THE LITERATURE

The material of Sections 3.2–3.5 (i.e. the applicaton of state-space methods to the optimization of LQ systems) is due to a great number of authors. It is now very standard, and to be found treated extensively in a number of general texts (e.g. Bertsekas, 1976; Whittle, 1982). We collect references on the Hamiltonian material of Section 3.7 at the end of Chapter 10.

The numerical methods of value iteration and policy improvement mentioned in Section 3.8 were originally formalized by Howard (1960) and are covered in any standard text (e.g. Bertsekas, 1976; Whittle, 1982; Ross, 1983). The fact that, in the LQ context, policy improvement is equivalent to the application of the Newton–Raphson algorithm to the Riccati equation was demonstrated by Whittle and Komarova (1988); see also Whittle (1989a).

CHAPTER 4

The Markov case: state estimation

4.1 IMPERFECT OBSERVATION AND CERTAINTY EQUIVALENCE

Imperfect observation will, in general, imply not only that one observes less than one might but also that observations are error-laden. The error is usually represented by a random variable: additive 'observation noise'. Observation is then a matter of degree—what is unobserved is observed with infinite error (i.e. with error of infinite variance).

A treatment of optimal control with imperfect observation thus forces us to a full statistical model, allowing both noise-corrupted observations and noise-disturbed plant equation. We thus modify assumptions (M1–4) of Section 3.2 as follows:

(M1)' As (M1): The state variable x and control variable u take values in R^n and R^m, respectively, and may take any values in these spaces.

(M2)' The plant equation has the linear form

$$x_t = Ax_{t-1} + Bu_{t-1} + \alpha_t + \varepsilon_t \tag{4.1}$$

where $\{\alpha_t\}$ is a known sequence and $\{\varepsilon_t\}$ is white noise, specified more closely in (M4)'.

(M3)' As (M3): operation is over the time interval $[0, h]$ with the cost function specified in equations (3.8)–(3.10).

(M4)' At time $t = 0$ one is given an estimate $\hat{x}_0$ of x_0 whose error $\Delta_0 = \hat{x}_0 - x_0$ is normally distributed with zero mean and covariance matrix V_0. At time t $(t > 0)$ one observes the quantity

$$y_t = Cx_{t-1} + \eta_t \tag{4.2}$$

where $\{\varepsilon_t, \eta_t; t > 0\}$ form a Gaussian white-noise process with zero mean and covariance matrix

$$\text{cov}\begin{bmatrix} \varepsilon \\ \eta \end{bmatrix} = \begin{bmatrix} N & L \\ L' & M \end{bmatrix} \tag{4.3}$$

this process being independent of Δ_0.

Let us write the optimal control (3.26) for the full-information noiseless case as

$$u_t = K_t x_t + \sum_{j=1}^{h-t} G_{tj} \alpha_{t+j} \tag{4.4}$$

Let us also denote $x_t^{(t)}$, the MLE of x_t based upon $W_t = (\hat{x}_0, Y_t, U_{t-1})$, by $\hat{x}_t$. Then an application of the certainty-equivalence principle yields.

Theorem 4.1.1. *Under assumptions* (M1–4)' *the optimal control takes the form*

$$u_t = K_t \hat{x}_t + \sum_{j=1}^{h-t} G_{tj} \alpha_{t+j} \qquad (4.5)$$

where the matrix coefficients K, G are those of the optimal full-information control. In particular, the optimal control in the case $\alpha = 0$ is simply

$$u_t = K_t \hat{x}_t \qquad (4.6)$$

Proof. All the assumptions required for the establishment of certainty equivalence in Theorem 2.5.1 are satisfied, with the noise vector ξ identified as $(\Delta_0; \varepsilon_t, \varepsilon_2, \ldots, \varepsilon_h; \eta_1, \eta_2, \ldots, \eta_h)$. If ξ were fully observable then the optimal control would be

$$u_t = K_t x_t + \sum_j G_{tj}(\alpha_{t+j} + \varepsilon_{t+j}) \qquad (4.7)$$

(Note an implied assumption: that full observation of ξ implies full observation of x. This goes back to the point made at the beginning of the section: one should regard the observations (4.2) (which might not determine x even if ε and η were known) as being formally supplemented by further observations of infinite error variance.)

We can now appeal to the certainty-equivalence principle in the unreduced form stated in Section 2.6, and replace all unobservables in (4.7) by their current MLEs. This means replacing x_t by $x_t^{(t)} = \hat{x}_t$ and ε_{t+j} by $\varepsilon_{t+j}^{(t)}$. The white-noise property of ε implies that $\varepsilon_{t+j}^{(t)} = 0$ ($j > 0$). □

If we regard the matrices K_t and G_{tj} as having been determined by the techniques of Chapter 3, then determination of the control (4.5) is complete, but for the calculation of the estimate of current state $\hat{x}_t$. It is at this point that we meet the celebrated Kalman filter.

4.1.1 Exercise and comments

(1) If we multiply equation (4.2) by LM^{-1} and subtract it from (4.1) then we obtain a form of the plant equation in which, if transformed quantities are indicated by a dagger, then

$$A^\dagger = A - LM^{-1}C \qquad B^\dagger = B$$
$$\alpha_t^\dagger = \alpha_t + LM^{-1}y_t$$
$$\varepsilon_t^\dagger = \varepsilon_t - LM^{-1}\eta_t$$

and thus $L^\dagger = 0$, $N^\dagger = N - LM^{-1}L'$, $M^\dagger = M$. The effect of the transformation is thus to normalize L to zero, just as S was normalized to zero in the control context (c.f. Exercise 3.2.2).

4.2 RECURSIVE STATE ESTIMATION: THE KALMAN FILTER

Theorem 4.2.1. *Under assumptions* (M1–4)′ *of Section* 4.1 *the distribution of* x_t *conditional on knowledge of* $W_t = (\hat{x}_0, Y_t, U_{t-1})$ *is normal with mean* $\hat{x}_t$ *and covariance matrix* V_t, *where these quantities are determined recursively by*

$$\hat{x}_t = A\hat{x}_{t-1} + Bu_{t-1} + \alpha_t + H_t(y_t - C\hat{x}_{t-1}) \tag{4.8}$$

$$V_t = pV_{t-1} \tag{4.9}$$

Here

$$H_t = (L + AV_{t-1}C')(M + CV_{t-1}C')^{-1} \tag{4.10}$$

and p is an operator having the effect

$$pV = N + AVA' - (L + AVC')(M + CVC')^{-1}(L' + CVA') \tag{4.11}$$

Initial values are provided by the prescribed $\hat{x}_0$ *and* V_0.

Equation (4.8) is the *Kalman filter*, a recursive relation for $\hat{x}_t$ which has the attractive form of the plant equation, with state value x_t replaced by the state estimate $\hat{x}_t$ and plant noise ε_t replaced by a term proportional to the *innovation* $y_t - C\hat{x}_{t-1}$. This quantity actually is the innovation, as defined in Section 2.4, because $C\hat{x}_{t-1} = y_t^{(t-1)}$, and so $y_t - C\hat{x}_{t-1}$ represents the component of y_t not predictable from information available at time $t - 1$.

The conclusions of the theorem are indeed due to Kalman (Kalman, 1960, 1963; Kalman and Bucy, 1961), although derived earlier in a non-dynamic context by Plackett (1950). There are many proofs of the result; in addition to that we now give we indicate others in Exercise 4.2.2, Exercise 5.3.3 and Section 4.4.

Proof. There is in fact no appeal to condition (M3)′; the form of the cost function is irrelevant. The distribution of x_t conditional on W_t will necessarily be normal, since x_t is a linear combination of control and α terms whose value is known at time t and random variables which are jointly normally distributed with the observations. If one indeed denotes the conditional mean and covariance matrix by $\hat{x}_t$ and V_t then the estimation error

$$\Delta_t = \hat{x}_t - x_t \tag{4.12}$$

is necessarily independent of W_t.

Consider now the moment when u_{t-1} has been determined but y_t not yet observed. We see from the plant and observation relations (4.1) and (4.2) that the distribution of x_t and y_t conditional on (W_{t-1}, u_{t-1}) is jointly normal with expectations

$$E(x_t | W_{t-1}, u_{t-1}) = A\hat{x}_{t-1} + Bu_{t-1} + \alpha_t$$
$$E(y_t | W_{t-1}, u_{t-1}) = C\hat{x}_{t-1} \qquad (4.13)$$

Let us denote these values by x_t° and y_t°. It follows then from (4.13) and the plant/observation relations that $x_t - x_t^\circ = \varepsilon_t - A\Delta_{t-1}$ and $y_t - y_t^\circ = \eta_t - C\Delta_{t-1}$. Thus, conditional on (W_{t-1}, u_{t-1}), these quantities are distributed normally with zero means and covariance matrix

$$\text{cov}\begin{bmatrix} \varepsilon_t - A\Delta_{t-1} \\ \eta_t - C\Delta_{t-1} \end{bmatrix} = \begin{bmatrix} N + AV_{t-1}A' & L + AV_{t-1}C' \\ L' + CA_{t-1}V' & M + CV_{t-1}C' \end{bmatrix} = \begin{bmatrix} V_{xx} & V_{xy} \\ V_{yx} & V_{yy} \end{bmatrix} \qquad (4.14)$$

say. Here we have appealed to the fact that Δ_{t-1} is a function of process/observation history before time t, and so independent of ε_t and η_t.

Now, if we calculate the distribution of x_t conditional on $(W_{t-1}, u_{t-1}, y_t) = W_t$, then we obtain exactly the conditional distribution we require. However, we know that the distribution of $x_t - x_t^\circ$ conditional on the value of $y_t - y_t^\circ$ (and of W_{t-1}, u_{t-1}) is normal with mean $V_{xy} V_{yy}^{-1} (y_t - y_t^\circ) = H_t(y_t - C\hat{x}_{t-1})$ and covariance matrix $V_{xx} - V_{xy} V_{yy}^{-1} V_{yx} = pV_{t-1}$. That is, $\hat{x}_t$ and V_t are evaluated as asserted in (4.8)–(4.11). □

Note that equation (4.9) is again a Riccati equation, an analogue of (3.13), but differing in that it is a forward rather than a backward recursion, and matrices are replaced by transposes of their analogues. The analogue of the determination (3.15) of the control matrix K_t is the determination (4.10) of the innovation coefficient matrix H_t. In fact, the correspondence between the cases of noiseless control optimization and policy-unspecified state estimation is complete, and provides one of the many 'dualities' which arise. In anticipation of this we have written the two Riccati recursions as $V_t = pV_{t-1}$ and $\Pi_t = f\Pi_{t+1}$, p and f standing for 'past' and 'future', respectively.

Somewhat less evidently, the Kalman filter (4.8) is the analogue of the recursion (3.25) determining σ_t in the control optimization of a deterministically disturbed plant.

4.2.1 Exercises and comments

(1) Note that the recursions (4.8)–(4.11) still hold if the Gaussian hypotheses are dropped and one interprets $\hat{x}_t$ simply as $\mathscr{E}(x_t | W_t)$: the linear least square estimate of x_t based upon W_t.

(2) The form of the Kalman filter strikes everyone at first encounter by the elegance of its interpretation: a plant equation driven by the innovation $\zeta_t = y_t - C\hat{x}_{t-1}$ instead of plant noise. However, this form is inevitable. Even in the

LS approach (i.e. without postulating normality) we necessarily have

$$\mathscr{E}(x_t|W_t) = \mathscr{E}(x_t|W_{t-1}, u_{t-1}) + H\zeta_t$$

for some H (c.f. equation (2.26)), and the estimate $\mathscr{E}(x_t|W_{t-1}, u_{t-1})$ will give the component of forecast $A\hat{x}_{t-1} + Bu_{t-1} + \alpha_t$ derived from the plant equation. In fact, H necessarily equals $\operatorname{cov}(x_t, \zeta_t)\operatorname{cov}(\zeta_t)^{-1}$, agreeing with evaluation (4.10).

(3) For the control optimization of Chapter 3 we imposed the condition $Q > 0$, in order to ensure uniqueness and finiteness of the optimal control. The analogous assumption in the present case would now be that $M > 0$. That is, that no observation is exact, which seems a strange requirement.

However, if it were true that $M\beta = 0$ for some vector β, then this would imply that $\beta'\eta$ was zero in mean square, and so that $L'\beta = 0$, which makes LM^{-1} well defined. Correspondingly, all expressions in (4.10) and (4.11) are well defined, even if $M + CVC'$ is singular. These points resolve themselves in the formulations of Chapter 5 and Part III.

(4) The analogue of the gain matrix $\Gamma_t = A + BK_t$ is $\Omega_t = A - H_t C$. Confirm that

$$\Delta_t = \Omega_t \Delta_{t-1} + H_t \eta_t - \varepsilon_t$$

Therefore for Ω_t to have a limit value (as t increases) which is a stability matrix ensures that $V_t = \operatorname{cov}(\Delta_t)$ also has a finite limit value.

(5) Note that the Kalman filter can be written

$$\hat{x}_t = \Omega_t \hat{x}_{t-1} + (Bu_{t-1} + H_t y_t + \alpha_t)$$

in which form the analogy with equation (3.25) is more evident.

(6) At the end of the proof of Theorem 4.2.1 we appealed to some standard identities for conditional normal distributions. In the control/estimation duality these identities are the duals of the identities for partial minimization of a quadratic form noted in Exercise 3.2.1.

(7) Suppose that one indeed has perfect observation in that (4.2) reduces to $y_t = x_{t-1}$. Thus $C = I$ and $L = 0$, $M = 0$, but one still has a unit time lag in observation. Show that $H_t = A$ and equations (4.8) and (4.9) have the solutions $V_t = N$, $\hat{x}_t = Ax_{t-1} + Bu_{t-1} + \alpha_t$ ($t > 0$).

4.3 OBSERVABILITY ETC.: INFINITE-HISTORY LIMITS

The analogue of the deviation-sensitivity condition of Section 3.4 would be the condition that, with L normalized to zero, N should be positive definite on $\{(A')^t\}$. That is, that

$$\sum_{t=0}^{r-1} A^t N (A')^t > 0 \qquad (4.15)$$

for some r. Again, either there is no such r or there is a value $r \leq n$. One might regard (4.15) as a condition of *excitability*: that in the purely noise-driven plant equation

$$x_t = Ax_{t-1} + \varepsilon_t$$

x is so stimulated by the noise that after a sufficient time its distribution is not confined to any subspace of R^n. In the general case one must replace A and N by the normalized values $A^\dagger$ and $N^\dagger$ defined in Exercise 4.1.1.

The analogue of the condition of controllability is that of *observability*. This condition requires that, if we consider the system (4.1), (4.2) in the case of zero noise and known control values, then there exists an r such that observation of $y_1, y_2, \ldots, y_r$ will determine x_υ (and so all subsequent x_t). This condition is equivalent to the condition that the matrix

$$\begin{bmatrix} C \\ CA \\ CA^2 \\ CA^{r-1} \end{bmatrix}$$

should have rank n, for some r. This is, in turn, equivalent to the requirement that, for an $M > 0$, the matrix $C'M^{-1}C$ should be positive definite on $\{A^t\}$.

The analogue of the condition of stabilizability is that of *detectability*: that there should exist a matrix H such that $A - HC$ is a stability matrix. The point of this condition is that, if we consider an estimate generated by a Kalman filter, but using this possibly non-optimal fixed H, then the estimate will have an error whose covariance matrix will tend to a finite limit with increasing t if $A - HC$ is a stability matrix (c.f. Exercise 4.2.4).

The analogue of a horizon which recedes to infinity is now a starting point which is indefinitely remote, which corresponds just to increasing t. The analogue of Theorem 4.1.1 is immediate.

Theorem 4.3.1. *Suppose the system observable, excitable and with noisy observation in that $M > 0$. Then*
(i) *The equilibrium Riccati equation*

$$V = pV$$

has a unique non-negative definite solution V.
(ii) $V_t \to V$ *as t increases for any finite non-negative definite V_0.*
(iii) *The limit matrix*

$$\Omega = A - HC = A - (L + AVC')(M + CVC')^{-1}C$$

is a stability matrix.

These assumptions are grossly excessive. For example, if M is singular then this means that there are components of observation noise which are absent, and this should make the task of state estimation easier rather than otherwise. However, the assumptions give a way of ensuring regularity in the sense that Ω is uniquely defined and fully stable. Results generally have a clear limit version in the case when M tends to singularity (see Exercise 4.2.3 and Section 4.5).

4.4 ALTERNATIVE FORMS

As in Section 3.5, we can deduce alternative expressions for pV and H_t.

Theorem 4.4.1. *The expressions* (4.9) *and* (4.10) *for* H_t *and* pV *can be alternatively written*

$$H_t = LM^{-1} + A^\dagger (C'M^{-1}C + V_{t-1}^{-1})^{-1} C'M^{-1} \tag{4.16}$$

$$pV = N^\dagger + A^\dagger (C'M^{-1}C + V_{t-1}^{-1})^{-1} (A^\dagger)' \tag{4.17}$$

Here $A^\dagger = A - LM^{-1}C$ and $N^\dagger = N - LM^{-1}L'$ are the values to which A and N reduce under the normalization of L to zero.

The matrix $C'M^{-1}C$ is the Fisher *information matrix*, the information that a single observation y_t conveys on x_{t-1}. It plays the same role in this context as does the control-power matrix $BQ^{-1}B'$ in the control context.

Singularity of either M or V is quite permissible: the matrix $(C'M^{-1}C + V^{-1})^{-1}$ is then well defined, although possibly also singular.

Proof. Let us, for simplicity, assume L normalized to zero. Then the joint probability density of x_{t-1}, x_t and y_t conditional on W_{t-1} and u_{t-1} is proportional to $\exp(-\frac{1}{2}\mathbb{D})$, where

$$\begin{aligned}\mathbb{D} &= [(x - \hat{x})' V^{-1} (x - \hat{x})]_{t-1} \\ &+ (x_t - Ax_{t-1} - Bu_{t-1} - \alpha_t)' N^{-1} (x_t - Ax_{t-1} - Bu_{t-1} - \alpha_t) \\ &+ (y_t - Cx_{t-1})' M^{-1} (y_t - Cx_{t-1})\end{aligned}$$

Now, to integrate out x_{t-1} is equivalent to minimizing it out in $\mathbb{D}$; we then have $\hat{x}_t$ and V_t determined by

$$\min_{x_{t-1}} \mathbb{D} = [(x - \hat{x})' V^{-1} x - \hat{x})]_t + \cdots$$

where $+ \cdots$ represents terms independent of x_t. We would then have

$$\min_{x_t} \min_{x_{t-1}} [\mathbb{D} - 2l' x_t] = -l' V_t l - 2l' \hat{x}_t + \cdots \tag{4.18}$$

where $+ \cdots$ now represents terms independent of l. If we perform the minimizations in (4.18) in reverse order we obtain evaluations of V_t and $\hat{x}_t$ which imply the evaluations (4.16) and (4.17) in the reduced case $L = 0$. Evaluations (4.16) and (4.17) then follow for the unreduced case by inversion of the normalizing transformations of Exercise 4.1.1. □

66   THE MARKOV CASE: STATE ESTIMATION

This proof foreshadows the material of Chapter 5. The estimate $\hat{x}_t$ is characterized as an MLE rather than an LSE; the inverse matrices in the exponent of the probability density are then cleared by introduction of the conjugate variable l.

4.5 CONTINUOUS TIME

If we wish to derive the formal continuous-time limit by supposing time to advance in increments of δ and then letting δ tend to zero we must substitute for A, B, α, N, M and L the quantities $I + A\delta$, $B\delta$, $\alpha\delta$, $N\delta$, $M\delta^{-1}$ and L. The plant and observation relations then become

$$\dot{x} = Ax + Bu + \alpha + \varepsilon$$
$$y = Cx + \eta \tag{4.19}$$

where (ε, η) is a white-noise process with power matrix (4.3). Note the differing role of δ in the transformation of N, M and L. This is because of the differential and non-differential nature of plant and observation relations, respectively.

As in Theorem 4.2.1, we find the law of $x(t)$ conditional on $W(t)$ to be normal with mean $\hat{x}(t)$ and covariance matrix $V(t)$ (for $t \geq 0$ if true for $t = 0$). The Kalman filter equation now becomes

$$\dot{\hat{x}} = A\hat{x} + Bu + \alpha + H(y - C\hat{x}) \tag{4.20}$$

where

$$H = (L + VC')M^{-1} \tag{4.21}$$

and $V = V(t)$ obeys the Riccati equation

$$\dot{V} = N + AV + VA' - (L + VC')M^{-1}(L' + CV) \tag{4.22}$$

or

$$\dot{V} = N^\dagger + A^\dagger V + V(A^\dagger)' - VC'M^{-1}CV \tag{4.23}$$

Note again the appearance of the information matrix $C'M^{-1}C$. If M is singular then $C'M^{-1}C$ is infinite (we assume C of full row-rank). In this case one is receiving perfect information on some aspect of x, and V must move instantaneously onto a locus for which $VC'M^{-1}CV$ is finite.

Thus suppose that a lies in the null space of M, so that $Ma = 0$. Then $a'y = a'Cx$ is observed perfectly. If $VC'M^{-1}CV$ is to be finite then VC' must be orthogonal to all elements of the null space of M. Thus $VC'a = 0$ and so $a'CVC'a = 0$. This implies that $a'Cx$ is estimated with zero error, as is clearly the case.

CHAPTER 5

The complete dualization of past and future

5.1 CONTROL/ESTIMATION COMPARISONS; PATH INTEGRALS

Although the calculation of the least-square estimate $\hat{x}_t$ in Section 4.2 led to a Riccati equation $V_t = pV_{t-1}$ in direct analogue to the Riccati equation $\Pi_t = f\Pi_{t+1}$ of control optimization, the two problems have yet to be brought into complete correspondence. The optimal control was derived by minimizing a quadratic form integrated over the future ($\tau \geq t$), the cost $\mathbb{C}_t$ from time t onwards. If future decisions are optimized for known x_t then the minimal cost is the quadratic function $[x'\Pi x - 2\sigma' x]_t + \cdots$, where $+ \cdots$ represents terms independent of x_t.

Determination of the optimal estimate $\hat{x}_t$ can also be represented in terms of minimization of a quadratic form, since it can be regarded as the maximum likelihood estimate. That is, if we denote the quadratic form in the exponent of the Gaussian density $f(X_t, Y_t | U_{t-1})$ by $-\frac{1}{2}\mathbb{D}_t$ then $\mathbb{D}_t$ is a quadratic form integrated over the past ($\tau \leq t$), and one obtains the estimates of the state variable $x_\tau^{(t)}$ over the history of the process based on information at time t by minimizing $\mathbb{D}_t$ with respect to $x_\tau(\tau \leq t)$. Alternatively, if we minimize only with respect to past unobservables $x_\tau(\tau < t)$ then we are left with an expression $[(x - \hat{x})V^{-1}(x - \hat{x})]_t + \cdots$. The minimization is equivalent to integrating these variables out in the density and so to calculating $f(x_t | W_t)$; see Exercise 5.1.1.

The analogy between the two optimizations is that already emphasized in Section 2.6. However, notice the difference. The matrices Π and V seem to be mutual analogues, but the quadratic form in x_t derived by minimizing $\mathbb{C}_t$ with respect to future decisions has Π_t as matrix, while the quadratic form in x_t derived by minimizing $\mathbb{D}_t$ with respect to past unobservables has the *inverse* V_t^{-1} as matrix. There is thus a discrepancy, accentuated by the fact that V_t may even be singular.

The discrepancy is yet further accentuated if we look at the path integrals themselves. The form for the cost function $\mathbb{C}_t$ is that assumed in Section 3.2, namely

$$\mathbb{C}_t = \sum_{\tau=t}^{h-1} \left[\begin{bmatrix} x \\ u \end{bmatrix}' \begin{bmatrix} R & S' \\ S & Q \end{bmatrix} \begin{bmatrix} x \\ u \end{bmatrix} \right]_\tau + (x'\Pi x)_h \tag{5.1}$$

in the homogeneous case, involving no matrix inverses. The form for $\mathbb{D}_t$ is

$$\mathbb{D}_t = (\Delta' V^{-1} \Delta)_0 + \sum_{\tau=1}^{t} \left[\begin{bmatrix} \varepsilon \\ \eta \end{bmatrix}' \begin{bmatrix} N & L \\ L' & M \end{bmatrix}^{-1} \begin{bmatrix} \varepsilon \\ \eta \end{bmatrix} \right]_\tau \tag{5.2}$$

Here Δ, ε and η are to be understood as expressed in terms of x, y and u by appeal to the plant and observation relations:

$$\Delta_0 = \hat{x}_0 - x_0 \tag{5.3}$$

$$\varepsilon_\tau = x_\tau - Ax_{\tau-1} - Bu_{\tau-1} - \alpha_\tau \tag{5.4}$$

$$\eta_\tau = y_\tau - Cx_{\tau-1} \tag{5.5}$$

Again, in expression (5.2) the matrices which seem to be analogous to the matrices in (5.1) occur as *inverses* rather than directly. Furthermore, some of the matrices inverted may well be singular.

We shall bring the two problems into complete correspondence by a transformation of the estimation problem, a transformation whose inevitability emerges even more clearly in the later LEQG and Hamiltonian formulations.

Note, in passing, that one could indeed calculate $\hat{x}_t = x_t^{(t)}$ by simultaneously minimizing $\mathbb{D}_t$ with respect to all $x_\tau (\tau \le t)$ and so re-estimating the whole history at time t. Perhaps one does not wish to do so, and the recursive techniques of the Kalman filter and the Hamiltonian techniques we shall outline in Section 5.3 are ways of estimating current state without simultaneously re-estimating the whole process history. Nevertheless, there is virtue in at least being aware that, behind the estimation of x_t, there lies an implicit estimation of the whole path history, just as the certainty-equivalence principle of Section 2.5 showed that behind the optimization of u_t there is an implicit optimization also of future actions.

We achieve the complete dualization of the two problems in Section 5.2, and outline the estimation version of the Hamiltonian formulation in Section 5.3. This material is original, but cannot be totally new—in the literature there must be many foreshadowings and partial versions of the approach taken.

5.1.1 Exercises and comments

(1) A key property of the multivariate normal distribution is that elimination of a variable by integration is equivalent (to within evaluation of a normalizing constant) to elimination by maximization. Explicitly, suppose vectors x and y are jointly normal with density

$$f(x, y) \propto \exp[-\tfrac{1}{2} Q(x, y)]$$

where Q is a quadratic form. Then

$$\int f(x,y)dx \propto \exp\left[-\frac{1}{2}\min_x Q(x,y)\right] \quad (5.6)$$

where the proportionality constant is independent of y (see Lemma 6.1.1).
For this reason we have

$$\min_{X_{t-1}}[-2\log f(X_t, Y_t|U_{t-1})] = [(x-\hat{x})V^{-1}(x-\hat{x})]_t + \cdots, \quad (5.7)$$

a relation on which the recursions for $\hat{x}_t$ and V_t can be based.

(2) We could have taken the complete path integral $\mathbb{D}_h$ and minimized with respect to *all* unobservables at time t, so deriving predictions of future x and y as well as estimates of past and present x. The estimates $x_\tau^{(t)}(\tau \leq t)$ of the latter will be the same as those obtained by minimization of $\mathbb{D}_t$; the estimates of the former essentially follow from $\varepsilon_\tau^{(t)} = 0$ and $\eta_\tau^{(t)} = 0$ ($\tau \geq t$). That is, the recursions

$$\begin{aligned} x_\tau^{(t)} &= Ax_{\tau-1}^{(t)} + Bu_{\tau-1} + \alpha_\tau \\ y_\tau^{(t)} &= Cx_{\tau-1}^{(t)} \end{aligned} \quad (5.8)$$

hold for $\tau > t$.

(3) Recursions (5.8) were derived on the assumption that future control values u_τ were prescribed in advance. Show that, if they are determined by a policy which prescribes u_τ as a function $u_\tau(\hat{x}_0, Y_\tau, U_{\tau-1})$ *linear* in the bracketed variables, then relations (5.8) still hold, with

$$u_\tau^{(t)} = u_\tau(\hat{x}_0, Y_\tau^{(t)}, U_{\tau-1}^{(t)})$$

substituted for u_τ.

5.2 INTRODUCTION OF THE CONJUGATE VARIABLES

Consider the minimization of $\mathbb{D}_t$ as the minimization of the form (5.2) in the variables x, Δ, ε and η subject to constraints (5.3)–(5.5). Let us introduce Lagrange multipliers l_τ, m_τ for these constraints, so that we consider the Lagrangian form

$$\mathbb{D}_t + l_0'(x_0 - \hat{x}_0 + \Delta_0) + \sum_{\tau=1}^t l_\tau'(x_\tau - Ax_{\tau-1} - Bu_{\tau-1} - \alpha_\tau - \varepsilon_\tau)$$

$$+ \sum_{\tau=1}^t m_\tau'(y_\tau - Cx_{\tau-1} - \eta_\tau) \quad (5.9)$$

where $\mathbb{D}_t$ has the expression (5.2). We should properly write the multipliers as $l_\tau^{(t)}, m_\tau^{(t)}$ to emphasize their dependence upon t. However, we shall include the t superscript where emphasis is needed and, for simplicity, neglect it otherwise.

Theorem 5.2.1. *If one minimizes out the $(\Delta, \varepsilon, \eta)$ variables in (5.9) then one reduces the problem to that of rendering the path integral*

$$\mathbb{I}(x, l, m) = [l'Vl - 2l'(x - \hat{x})]_0 + \sum_{\tau=1}^{t} [v(l_\tau, m_\tau)$$

$$- 2l'_\tau(x_\tau - Ax_{\tau-1} - Bu_{\tau-1} - \alpha_\tau) - 2m'_\tau(y_\tau - Cx_{\tau-1})] \quad (5.10)$$

maximal with respect to the x variables and minimal with respect to the (l, m) variables. Here

$$v(l, m) = \begin{bmatrix} l \\ m \end{bmatrix}' \begin{bmatrix} N & L \\ L' & M \end{bmatrix} \begin{bmatrix} l \\ m \end{bmatrix} \quad (5.11)$$

and the following relations hold between multipliers and noise estimates:

$$V_0 l_0^{(t)} = -\Delta_0^{(t)} = x_0^{(t)} - \hat{x}_0 \quad (5.12)$$

$$\begin{bmatrix} N & L \\ L' & Q \end{bmatrix} \begin{bmatrix} l \\ m \end{bmatrix}_\tau^{(t)} = \begin{bmatrix} \varepsilon \\ \eta \end{bmatrix}_\tau^{(t)} \quad (5.13)$$

Proof. Equations (5.12) and (5.13) are just the stationarity conditions of the Lagrangian form with respect to the noise variables. When one eliminates the noise variables by means of these then one is left with the form $-\mathbb{I}(x, l, m)$, and remaining assertions then follow from the fact that x is a primal variable and l and m dual variables of the original constrained problem. □

Note that the effect of the transformation has been to clear the form of matrix inverses. The notation $v(l, m)$ for the quadratic form (5.11) is used in analogue to the notation $c(x, u)$ for the quadratic instantaneous cost (3.9). It is the form $\mathbb{I}$ rather than the form $\mathbb{D}$ which is the analogue of $\mathbb{C}$. In order to make this clear we shall develop the recursion which the extremized value of $\mathbb{I}$ satisfies.

Let us write stat$(\mathbb{I}|w)$ for an extremization of the path integral $\mathbb{I}$ (maximization or minimization, as appropriate for the variable) in which the variables w, a subset of the arguments of $\mathbb{I}$, are held fixed. So, by (5.7).

$$\text{stat}(\mathbb{I}|x_t) = \text{const.} - [(x - \hat{x})V^{-1}(x - \hat{x})]_t \quad (5.14)$$

Consequently

$$\text{stat}(\mathbb{I}|x_t, l_t) = \text{const.} + [l'Vl - 2l'(x - \hat{x})]_t \quad (5.15)$$

In consequence of (5.15) and the definition (5.10) of I we must have the recursion

$$[l'Vl - 2l'(x - \hat{x})]_t = \text{const.} + \underset{l_{t-1}, m_t, x_{t-1}}{\text{stat}} \{[l'Vl - 2l'(x - \hat{x})]_{t-1} + v(l_t, m_t)$$

$$- 2l'_t(x_t - Ax_{t-1} - Bu_{t-1} - \alpha_t) - 2m'_t(y_t - Cx_{t-1})\}$$

$$(5.16)$$

5.2 INTRODUCTION OF THE CONJUGATE VARIABLES

where the constant is a term independent of l_t or x_t. With this, we have the estimation problem in a form in which the complete duality with the control problem is apparent.

Theorem 5.2.2. *If one minimizes out x_{t-1} in recursion (5.15) then one obtains the relation*

$$l_{t-1} = A'l_t + C'm_t \tag{5.17}$$

(the analogue of the deterministic plant equation (3.7)), and the reduced recursion

$$(l'Vl + 2l'\hat{x})_t = \min_{m_t} \left[v(l_t, m_t) + (l'Vl + 2l'\hat{x})_{t-1} + 2l'_t(Bu_{t-1} + \alpha_t) - 2m'_t y_t \right]$$
$$+ \text{const.} \tag{5.18}$$

where l_{t-1} is to be substituted from (5.17).

Proof. Relation (5.17) is necessary if the extremum with respect to x_{t-1} in (5.15) is to be finite. Alternatively, extremization with respect to l_{t-1} and x_{t-1} (in that order) leads to the same conclusion. Remaining conclusions then follow directly. □

We see the parallel with the optimality equation for the disturbed regulation problem if we write this as

$$(x'\Pi x - 2\sigma' x)_t = \min_{u_t} \left[c(x_t, u_t) + (x'\Pi x - 2\sigma' x)_{t+1} \right] + \text{const.} \tag{5.19}$$

where x_{t+1} is to be substituted from

$$x_{t+1} = Ax_t + Bu_t + \alpha_{t+1} \tag{5.20}$$

We can express the parallel precisely.

Theoem 5.2.3. *Recursion (5.18) implies the recursions (4.8)–(4.11) for $\hat{x}_t$ and V_t. Insofar as the second-order terms in (l, m) are concerned it is the complete analogue of the control optimality equation (5.19) with the substitution of $(-l, -m, A', C', N, L', M, V)$ for $(x, u, A, B, R, S, Q, \Pi)$ and time reversal.*

The conclusions are immediate. It is also true that $\hat{x}_t$ is the analogue of σ_t, but this is less obvious, because the two problems, as they are set up, differ somewhat in their non-homogeneous terms. The point can be cleared, but the path-integral formulation of Part III gives an easy and general resolution of all these matters.

It would seem to be natural to change the sign in the definition of l and m. However, the convention we have taken will prove best.

Theorem 5.2.4. *The fuller recursion (5.15) yields also the alternative forms of the Kalman filter and Riccati equation for V asserted in Theorem 4.4.1 if the variables are minimized in the order l_{t-1}, m_t and then x_{t-1}.*

The assertion is easily verified. Indeed, the proof of Theorem 4.4.1 is a version of this calculation.

5.2.1 Exercises and comments

(1) The interpretation of l_τ and m_τ as Lagrange multipliers means that these can be interpreted as the rates of change of the minimized value of $\mathbb{D}_t$ with respect to a change (brought about by an exogenous input in the plant or observation relation) in x_τ resp. y_τ. The 'dual plant equation' (5.17) then makes sense: an enforced change of δ in $x_{\tau-1}$ would produce changes of $A\delta$ and $C\delta$ in x_τ and y_τ.

(2) The 'dual plant equation' (5.17) is a relation that holds whatever the value of l_t, in that l_t is being held as a free variable in (5.15). If in fact we write down the full set of stationarity conditions for $\mathbb{I}$ with respect to $x_\tau (\tau \leq t)$ we derive the rather more explicit equations

$$l_{\tau-1}^{(t)} = A'l_\tau^{(t)} + C'm_\tau^{(t)} \qquad (\tau \leq t) \tag{5.21}$$

with boundary condition

$$l_t^{(t)} = 0 \tag{5.22}$$

Relation (5.22) is to be construed as saying that giving x_t its most probable value is the same as giving ε_t its most probable value. This will be $\varepsilon_t = 0$ in the case $L = 0$, when one has no observational information on the value of ε_t at all.

5.3 THE HAMILTONIAN FORMALISM

If we write out the stationarity conditions with respect to $(l, m, x)_\tau (1 \leq \tau \leq t)$ for the path integral $\mathbb{I}(l, m, x)$ defined in (5.10) then we find that these can be written

$$\begin{bmatrix} N & L & \mathscr{A} \\ L' & M & \mathscr{C} \\ \bar{\mathscr{A}} & \bar{\mathscr{C}} & 0 \end{bmatrix} \begin{bmatrix} -l \\ -m \\ x \end{bmatrix}_\tau^{(t)} + \begin{bmatrix} \mathscr{B}u - \alpha \\ y \\ 0 \end{bmatrix}_\tau = 0 \qquad (1 \leq \tau \leq t) \tag{5.23}$$

Here we have again used the operator notation

$$\mathscr{A} = I - A\mathscr{T}, \qquad \mathscr{B} = -B\mathscr{T}, \qquad \mathscr{C} = -C\mathscr{T} \tag{5.24}$$

with

$$\bar{\mathscr{A}} = I - A'\mathscr{T}^{-1} \tag{5.25}$$

etc. Further, the translation operator $\mathscr{T}$ acts on the *subscript* τ in (5.23), not on the superscript t, which is regarded as fixed for the moment.

In equation (5.23) we see the complete formal dual of the corresponding system (3.52) for the control problem. The fact that this is indeed the complete dual confirms that the detailed analogue between past (estimation) and future (control) is achieved by the introduction of the conjugate variables l and m. We

5.3 THE HAMILTONIAN FORMALISM

see from (5.23) that it is the vector $(-l, -m, x)$ which is the analogue of the vector (x, u, λ) for the control case. Since λ and l are both Lagrange multipliers for the constraint implied by the plant equation, they should be related. The relation emerges in Section 10.2.

Equations (5.23) are subject to the boundary conditions

$$Vl_0 = x_0 - \hat{x}_0$$
$$l_0 = A'l_1 + C'm_1 \tag{5.26}$$

and

$$(l, m)_\tau = 0 \quad (\tau > t) \tag{5.27}$$

a superscript (t) being understood throughout.

If we write the system (5.23) as

$$\Psi(\mathcal{T})\chi_\tau^{(t)} = \rho_\tau \quad (1 \leq \tau \leq t) \tag{5.28}$$

then we see the same possibility of reducing the system by a canonical factorization as was achieved by the canonical factorization of Φ in Section 3.7. In particular, this factorization should, under regularity conditions, give an immediate solution of the infinite-history case, when observation began in the remote past rather than at $t = 0$. In this case system (5.28) will be replaced by the infinite system

$$\Psi(\mathcal{T})\chi_\tau^{(t)} = \rho_\tau \quad (\tau \leq t) \tag{5.29}$$

Assume that Ψ has the canonical factorization

$$\Psi = \Psi_+ \Psi_0 \Psi_- \tag{5.30}$$

analogous to equation (3.61) (although note the reversed order of factors). Then we expect that under regularity conditions it will be legitimate to partially invert (5.29) to

$$\Psi_-(\mathcal{T})\chi_\tau^{(t)} = \Psi_0^{-1}\Psi_+(\mathcal{T})^{-1}\rho_\tau \quad (\tau \leq t) \tag{5.31}$$

Equation (5.31) for $\tau = t$ will yield an explicit expression for $\hat{x}_t = x_t^{(t)}$ in terms of past observations $y_t, y_{t-1}, y_{t-2}, \ldots$. Indeed, if we rewrote relation (5.31) as separate backward and forward recursions

$$\Psi_-(\mathcal{T})\chi_\tau^{(t)} = v_\tau \tag{5.32}$$
$$\Psi_+(\mathcal{T})\Psi_0 v_\tau = \rho_\tau \tag{5.33}$$

then the forward recursion (5.33) is an analogue of the Kalman filter, determining the appropriate linear function v of past observations y. Relation (5.32) at $\tau = t$ then determines $\hat{x}_t$ directly in terms of v_t, and for $\tau = t - 1, t - 2, \ldots$ then recursively determines $x_{t-1}^{(t)}, x_{t-2}^{(t)}, \ldots$. Exercise 5.3.3.

74 THE COMPLETE DUALIZATION OF PAST AND FUTURE

As previously, the well-posedness of the infinite-history problem requires that Ψ possess such a canonical factorization, and that the partial inversion (5.31) be legitimate.

5.3.1 Exercises and comments

(1) The canonical factorization (5.30) is derived from a maximum likelihood characterization, and is *not* the same as the familiar canonical factorization of Wiener prediction theory, which is derived from a least-square characterization. Consider, for example, the prediction of y_{t+r} from $y_t, y_{t-1}, \ldots$, under the simplifying assumptions that u and α are identically zero and L is zero. The ML characterisation gives a prediction

$$y_{t+r}^{(t)} = CA^r \hat{x}_t \tag{5.34}$$

with $\hat{x}_t$ calculated from the Kalman filter. The Wiener LS characterization determines $y_{t+r}^{(t)}$ as a linear function of the observables, the coefficients in this linear function being determined by canonical factorization of the autocovariance-generating function of the y process:

$$M + C(I - Az)^{-1} N(I - A'z^{-1})^{-1} C' \tag{5.35}$$

The factorization of this rational function is not the same as the factorization (5.30) of $\Psi(z)$, which has the simpler polynomial form

$$\Psi(z) = G_{-1} z^{-1} + G_0 + G_1 z \tag{5.36}$$

The two procedures lead to the same final estimate, but are very different otherwise. Expression (5.36) is clearly much simpler to factorize than (5.35), and the validity of relations such as (5.34) is immediately evident in the ML approach and far from evident in the LS approach.

(2) Assume that V_t, H_t have the limit values V, H. Verify that

$$\Psi_+(z) = \bar{\Psi}_-(z) = \begin{bmatrix} V & H & I - Az \\ 0 & I & -Cz \\ I & 0 & 0 \end{bmatrix} \quad \Psi_0 = \begin{bmatrix} 0 & 0 & I \\ 0 & M + CVC' & 0 \\ I & 0 & -V \end{bmatrix} \tag{5.37}$$

constitute a canonical factorization (5.30) of Ψ (c.f. equations (3.64) and (3.65)).

(3) Confirm that the Ψ_+ of (5.37) has inverse

$$\Psi(\mathcal{T})_+^{-1} = \begin{bmatrix} I & 0 & 0 \\ 0 & I & C z \\ 0 & 0 & I \end{bmatrix} \begin{bmatrix} 0 & 0 & I \\ 0 & I & 0 \\ \mathcal{H}^{-1} & -\mathcal{H}^{-1} H & -\mathcal{H}^{-1} V \end{bmatrix}$$

where

$$\mathcal{H} = I - A\mathcal{T} + HC\mathcal{T} = I - \Omega\mathcal{T}$$

5.3 THE HAMILTONIAN FORMALISM

The first and third of relations (5.31) at $\tau = t$ then yield

$$-Vl_t + \hat{x}_t = \mathcal{H}^{-1}(Bu_{t-1} + Hy_t + \alpha_t), \qquad l_t = 0$$

with the implication that

$$\mathcal{H}\hat{x}_t = Bu_{t-1} + Hy_t + \alpha_t$$

This is just the limit form of the Kalman filter (4.8).

PART II

Risk sensitivity: the LEQG formulation

CHAPTER 6

The risk-sensitive certainty-equivalence principle

6.1 THE LEQG FORMULATION

The notion of risk sensitivity has already been discussed in Section 1.2. The natural way of achieving risk sensitivity in the LQG context is to retain all the LQG assumptions (LQG1–5) of Section 2.2, but to replace the minimization of the criterion $E_\pi(\mathbb{C})$ with respect to policy π by the minimization of the criterion

$$\gamma_\pi(\theta) = -(2/\theta)\log(E_\pi \exp(-\theta\mathbb{C}/2)) \tag{6.1}$$

Here $\mathbb{C}$ is, as ever, the quadratic cost function, and θ can be characterized as the risk-sensitivity parameter. Note that we are as yet making no assumptions of state structure. As explained in Section 6.2, increasingly positive values of θ correspond to an ever more optimistic (*risk-seeking*) attitude on the part of the optimizer; increasingly negative values correspond to an increasingly pessimistic (*risk-averse*) attitude. The case $\theta = 0$ is the *risk-neutral* case, reducing to the standard problem of minimizing $E_\pi(\mathbb{C})$.

We shall follow a convention already current and term this generalization of the LQG formulation the *LEQG formulation*.

One might justify the exponential change of utility scale by its decoupling of the effects of expected cost and cost variability; see Exercise 1.2.3. However, the overwhelming reason for the particular choice of exponential-of-quadratic as criterion function is that this fits in naturally with the form of the Gaussian distribution. The following Lemma, already mentioned in Exercise 5.1.1, makes this point.

Lemma 6.1.1. *Let $Q(x,y)$ be a positive-definite quadratic form in the vector variables x and y. Then*

$$\int \exp\left[-\frac{1}{2}Q(x,y)\right]dx \propto \exp\left[-\frac{1}{2}\min_x Q(x,y)\right] \tag{6.2}$$

(The variable remaining on both sides is y, so it is understood that the proportionality constant is independent of y.)

Proof. The identity follows from the usual 'completion of the square' evaluation of a normal integral. If $\hat{x}$ is the minimizing value of x in (6.2) then

$$Q(x, y) = Q(\hat{x}, y) + (x - \hat{x})'H(x - \hat{x}) \tag{6.3}$$

where H is a constant matrix, the matrix of second derivatives of Q with respect to x. Identity (6.2) thus holds with the constant of proportionality evaluated as $\int \exp(-\frac{1}{2}\Delta' H \Delta)d\Delta$. □

Therefore we have the great simplification that, for exponential-of-quadratic functions, integration can be replaced by extremization of the exponent, at least if the quadratic form has the right definiteness properties.

In fact, we shall have a combination of the operations of integrating out unobservables and optimizing control variables. It will be found that the optimization will imply that one may wish to either maximize or minimize an integral, depending on the sign of θ.

Lemma 6.1.2. *Let $Q(x, y, u)$ be a quadratic form in the variables indicated, positive definite in x. Then*

$$\max_{x} \int \exp\left[-\frac{1}{2}Q(x, y, u)\right]dx \propto \exp\left[-\frac{1}{2}\min_{u}\min_{x} Q(x, y, u)\right] \tag{6.4}$$

if $\min_x Q(x, y, u)$ is positive definite in u for given y, and

$$\min_{u} \int \exp\left[-\frac{1}{2}Q(x, y, u)\right]dx \propto \exp\left[-\frac{1}{2}\max_{u}\min_{x} Q(x, y, u)\right] \tag{6.5}$$

if $\min_x Q(x, y, u)$ is negative definite in u for given y. In both cases the extremizing value of u is the same on both sides of the relation.

The assertions follow as a corollary of Lemma 6.1.1.

6.1.1 Exercises and comments

(1) We shall later wish to be able to assert that the order of the extremizing operations in the right-hand members of (6.4) and (6.5) is immaterial. Let us for the moment neglect the variable y, and so assume that we are calculating the extremes of a non-homogeneous quadratic form

$$Q(x, u) = \begin{bmatrix} x \\ u \end{bmatrix}' Q \begin{bmatrix} x \\ u \end{bmatrix} + \cdots = \begin{bmatrix} x \\ u \end{bmatrix}' \begin{bmatrix} Q_{11} & Q_{12} \\ Q_{21} & Q_{22} \end{bmatrix} \begin{bmatrix} x \\ u \end{bmatrix} + \cdots$$

where $+\cdots$ indicates terms of less than second degree. In case (6.4) $Q(x, u)$ can be minimized with respect to x and u in either order if it is jointly positive definite in both variables, i.e. if $Q > 0$.

In case (6.5) we shall certainly require that $Q_{11} > 0$ and $Q_{22} - Q_{21}Q_{11}^{-1}Q_{12} < 0$ (this second matrix being the matrix of the form $\min_x Q(x, u)$). The condition for validity of interchange of the two operations can most economically be written

$$\begin{bmatrix} Q_{11} & iQ_{12} \\ -iQ_{21} & -Q_{22} \end{bmatrix} > 0 \tag{6.6}$$

We shall speak of this as a *saddle-point condition* with respect to x and u.

6.2 THE RISK-SENSITIVE CERTAINTY-EQUIVALENCE PRINCIPLE (RSCEP)

Let us then make the assumptions (LQG1–5) of Section 2.2. However, the optimization goal is now the minimization of the risk-sensitive criterion function $\gamma_\pi(\theta)$ of (6.1) rather than of $\gamma_\pi(0) = E_\pi(\mathbb{C})$.

The LQG assumptions imply that we can express the cost $\mathbb{C}$ as a quadratic function of control history U_{h-1} and noise vector ξ. Define also the quadratic form

$$\mathbb{D} = \xi' V^{-1} \xi \tag{6.7}$$

where, be it recalled, V is the exogenously specified covariance matrix of ξ. The form $\mathbb{D}$ is thus that occurring in the exponent of the ξ-density. Define now the *total stress*

$$\mathbb{S} = \mathbb{C} + \theta^{-1} \mathbb{D} \tag{6.8}$$

This is a combination which occurs spontaneously when one calculates the expectation in (6.1). $\mathbb{C}$ is the component of stress due to *cost* (i.e. to departures of the process and control variables from the path one would wish them to have) and $\mathbb{D}$ is the component due to implausibility or *discrepancy* (i.e. to the process and control variables' taking values which imply that ξ is large).

As in Sections 2.2 and 2.5, we take $\mathbb{C}$ and $\mathbb{D}$ in these reduced forms in order to simplify the proof of the CEP. Once the CEP has been proved then, as in Section 2.6, we apply it to the unreduced version, in which $\mathbb{C}$ is expressed as a function of system path (X_h, U_{h-1}) and $\mathbb{D}$ is identified with the quadratic form in $-2 \log f(X_h, Y_h | U_{h-1})$.

We shall use the term θ-*extremization* to denote an operation in which one minimizes when $\theta \geq 0$ and maximizes when $\theta < 0$. This operation will be denoted 'ext' (analogously to 'min' and 'inf'). The unqualified term *extremization* will refer simply to the forming of an extreme of some unspecified and possibly mixed character. In some cases, when we have a stationary point of mixed character we shall simply use 'stat' to denote the operation of evaluation at the appropriate stationary point.

Theorem 6.2.1. *The risk-sensitive certainty-equivalence principle (RSCEP). Assume conditions (LQG1–5) of Section 2.2, and assume that the policy π is to be* chosen to minimize the criterion $\gamma_\pi(\theta)$ of (6.1). Suppose also, in the case $\theta < 0$, that θ has a value such that the quadratic form $\theta\mathbb{S}$ obeys a saddle-point condition with respect to ξ and U_{h-1}. Then

(i) *The optimal value of u_t is determined by simultaneously minimizing $\mathbb{S}$ with respect to $u_t, u_{t+1}, \ldots, u_{h-1}$ and θ-extremizing it with respect to $y_{t+1}, y_{t+2}, \ldots y_h$. In words, one obtains an optimal current decision by minimizing stress with respect to all decisions currently unmade and θ-extremizing it with respect to all quantities currently unobservable.*

(ii) *The value function $G(W_t)$ of the problem is related to the partially extremized stress $\mathbb{S}_t$ by equation (6.13) below.*

This is a result which is rich in implication for what can now be done and for the changed view of the problem to which one is now forced. We shall pursue these in the next section; first, the proof.

Proof. Let the observations y_t be the reduced observations, obtained by eliminating the known effects of past controls and initial conditions. They are thus simply linear functions of ξ. We lose no generality if we assume that these are linearly independent functions of ξ, since this simply amounts to discarding that information in a new observation which is already known from previous observations. We also lose no generality if we make the formally convenient assumption: that the complete observation vector Y_h gives one complete information; i.e. determines ξ. This is because the last observation y_h becomes available only after the last decision u_{h-1} has been taken, and so cannot affect decisions.

By making the known linear transformation from ξ to Y_h we can write the stress as a quadratic function

$$\mathbb{S} = \mathbb{S}(Y_h, U_{h-1}) = \mathbb{S}(W_h) \tag{6.9}$$

of the observables at time h.

Define the quantity

$$G(W_t) = \underset{\pi}{\text{ext}}\left[-f(Y_t/U_{t-1})E_\pi \exp\left(-\frac{1}{2}\theta\mathbb{C}\right)\right] \quad (0 \le t \le h) \tag{6.10}$$

where $f(Y_t/U_{t-1})$ is the probability density (Gaussian and policy independent) of Y_t. Expression (6.10) is then the value function associated with recursive θ-extremization of $-E_\pi \exp(-\frac{1}{2}\theta\mathbb{C})$, modified by the factor $f(Y_t)$. In terms of this, the optimality equation becomes

$$G(W_t) = \underset{u_t}{\text{ext}} \int G(W_{t+1}) dy_{t+1} \quad (0 \le t < h) \tag{6.11}$$

with terminal condition

$$G(W_h) = -\exp\left[-\frac{1}{2}\theta(k_h + \mathbb{S}(W_h))\right] \tag{6.12}$$

Here k_h is a constant, independent of W_h. The extremizing value of u_t in (6.11) is the optimal value.

We shall now verify by induction that expression (6.12) generalizes to

$$G(W_t) = -\exp\left[-\frac{1}{2}\theta(k_t + \mathbb{S}_t(W_t))\right] \tag{6.13}$$

Here $\mathbb{S}_t(W_t)$ is a t-dependent quadratic function of $W_t = (Y_t, U_{t-1})$, obeying the recursion

$$\mathbb{S}_t(W_t) = \min_{u_t} \operatorname*{ext}_{y_{t+1}} \mathbb{S}_{t+1}(W_{t+1}) \qquad (0 \le t < h) \tag{6.14}$$

with terminal condition

$$\mathbb{S}_h(W_h) = \mathbb{S}(W_h) \tag{6.15}$$

The induction will also demonstrate that the minimizing value of u_t in (6.14) is the optimal value. This conclusion implies assertion (i) of the Theorem, for it implies that $\mathbb{S}_t(W_t)$ is the stress $\mathbb{S}$ minimized with respect to $u_t, u_{t+1}, \ldots$ and θ-extremized with respect to $t_{t+1}, y_{t+2}, \ldots$, and that the value of u_t thus determined is optimal.

Evaluation (6.12) begins the induction. Assume then (6.13) to be true at $t + 1$, so that the optimality equation (6.11) becomes

$$G(W_t) \propto \operatorname*{ext}_{u_t} - \int \exp\left[-\frac{1}{2}\theta \mathbb{S}_{t+1}(Y_{t+1}, U_t)\right] dy_{t+1}$$

with positive proportionality constant.

An application of Lemma 6.1.2 then demonstrates that $G(W_t)$ indeed has the evaluation (6.13), with $\mathbb{S}_t$ being related to $\mathbb{S}_{t+1}$ by (6.14). Moreover, the minimizing value of u_t in (6.14) is the optimal value.

The saddle-point condition on $\theta\mathbb{S}$ is required so that all extremal operations are proper and can be applied in any order (c.f. Exercise 6.1.1). The condition is expressed in terms of $\theta\mathbb{S}$ because this is the form occurring in the exponent. □

6.3 IMPLICATIONS OF THE RSCEP

In practice, the components $\mathbb{C}$ and $\mathbb{D}$ of stress will be given in their unreduced forms, as functions of the process/control path and of unreduced observations. The RSCEP still holds, however, in the verbal form stated in Theorem 6.2.1. That is, if one simultaneously minimizes stress with respect to all decisions currently unmade and θ-extremizes it with respect to all current unobservables, then the determination of current control u_t obtained in this way is optimal. The optimization of u_t is thus embedded in the provisional determination of a number of quantities. Let the determination of a variable w obtained in this way be denoted by $w^{(t)}$, as previously, where t is the current moment of time. Let us term this the *minimal stress estimate* (MSE) of w at t. (The term 'minimal' is not quite correct if θ is negative, but is convenient.) Therefore $x_\tau^{(t)}$ (any τ) is an optimal estimate of the process variable x_τ, also $u_\tau^{(t)}$ ($\tau \geq t$) is the optimal estimate at time t of the optimal decision to be made at time τ, and $y_\tau^{(t)}$ ($\tau > t$) is the optimal predictor of future observations.

The RSCEP does indeed reduce to the risk-neutral version of the CEP as θ tends to zero (from either side). As θ tends to zero then the weight of $\mathbb{D}$ relative to $\mathbb{C}$ in the total stress $\mathbb{C} + \theta^{-1}\mathbb{D}$ becomes overwhelming, and MS estimates of unobservables reduce to the values minimizing $\mathbb{D}$, i.e. to the ML estimates. The controls then minimize $\mathbb{C}$ with unobservables set equal to their ML estimates. This is indeed just the risk-neutral CEP of Theorem 2.5.1, as elaborated in Section 2.6.

The risk-neutral CEP had two features which made it clearly a certainty-equivalence principle:

(1) *Conversion to free form.* A minimization with respect to realizable *functions* $u_\tau(W_\tau)$ is replaced by a simple minimization with respect to *constants* u_τ. Specifically, the free minimization of $Q(U_{h-1}, \xi^{(t)})$ with respect to $u_t, u_{t-1}, \ldots$ automatically yields the evaluation of an optimal realizable closed-loop control rule $u_t(W_t)$.
(2) *Separation.* Optimization of estimation and control are separate, in that estimates are derived without reference to determination of a control rule, and the controls are determined as they would be in the full-information case, with substitution for unobservables of their estimates.

In the risk-sensitive case we certainly retain property (1): conversion to free form. That is, a minimization with respect to realizable functions $u_\tau(W_\tau)$ has been replaced by a free minimization/extremization of stress with respect to relevant decisions/unobservables.

However, we no longer have the property of separation in the simple form stated in (2) above. Determinations of optimal controls and of optimal estimates are intertwined in the simultaneous minimization/extremization of stress. Indeed, as we shall see, the form of the cost function affects estimates and the form of noise statistics affects control.

This fact is inevitable. If 'separation' is a meaningful concept at all, it must surface in another and possibly less evident form. This it does, as we shall see when we consider the state-structured case in the next section. If state structure is assumed then total stress can be decomposed at the current instant t into past stress and future stress. If one takes the current state value x_t as a pivotal quantity, in that one provisionally assumes it known, then extremization of past stress and of future stress can be decoupled, which is very close to saying that the optimization of estimation and of control can be decoupled. This must be the new characterization of the separation principle. By solving these two separate problems one obtains an evaluation of 'minimal' stress conditional on knowledge of x_t. By choosing x_t to extremize this total stress (so equating x_t to its MSE) one effectively recouples the two problems. In the risk-neutral version this last step is equivalent to the substitution of $\hat{x}_t$ for x_t in the optimal control rule. In the risk-sensitive case it implies the substitution for x_t of an estimate which depends on costs incurred over the whole path of the process as well as on past observations.

Finally, it may be asserted that the RSCEP *is* the statement of an operationally useful stochastic maximum principle which has been sought for years, and which can *only* be seen by leaving the risk-neutral case for the more general risk-sensitive case. We justify this assertion in Section 10.4.

6.3.1 Exercises and comments

(1) Consider the problem with Markov plant equation

$$x_t = Ax_{t-1} + Bu_{t-1} + \varepsilon_t \tag{6.16}$$

cost function (3.8) and perfect state observation, Here ε is white noise with $\text{cov}(\varepsilon) = N$. The RSCEP then tells us that we should determine the optimal u_0 by minimizing

$$\mathbb{S} = \mathbb{C} + \theta^{-1} \sum_{1}^{h} (\varepsilon' N^{-1} \varepsilon)_t \tag{6.17}$$

with respect to $u_0, u_1, \ldots, u_{h-1}$ and θ-extremizing it with respect to $x_1, x_2, \ldots, x_h$, subject to (6.16). Equivalently, one can assume that one is minimizing with respect to the u-variables and θ-extremizing with respect to the ε-variables. However, the situation is then effectively one in which at time $t - 1$ one has *two* components of control available, u_{t-1} and ε_t. These enter the plant equation together, as we see from (6.16), and the ε-control incurs an immediate effective cost $\theta^{-1} \varepsilon'_t N^{-1} \varepsilon_t$, as we see from (6.17). We might regard u and ε as 'the optimizer's control' and 'Nature's control', respectively. If θ is positive then Nature is benign, in that she tries to make $\mathbb{C}$ small as well as make $\mathbb{D}$ small. That is, she tries to help the optimizer. In the case of negative θ Nature is malign, in that she tries to make $\mathbb{C}$ large as well as make $\mathbb{D}$ small. That is, she obstructs the optimizer.

Otherwise expressed, the optimizer effectively regards the unpredictable noise process ε as taking values to his advantage if θ is positive and to his disadvantage if θ is negative.

(2) Consider the uncontrolled, scalar infinite-horizon version of problem (6.16). We thus make the specializations $B = 0$ and $\mathbb{C} = \sum_0^\infty Rx_t^2$. In choosing $x_t(t > 0)$ to extremize

$$\mathbb{S} = \sum_0^\infty (Rx_t^2 + (\theta N)^{-1}(x_t - Ax_{t-1})^2) \tag{6.18}$$

for given x_0 we then have a risk-sensitive formulation of a pure prediction problem from time $t = 0$. In the risk-neutral case the ML prediction is evidently

$E(x_t|x_0) = A^t x_0$. The values $x_t^{(0)}$ extremizing stress satisfy

$$(1 + A^2 + \theta NR)x_t^{(0)} - A(x_{t-1}^{(0)} + x_{t+1}^{(0)}) = 0 \qquad (t > 0)$$

Assume, for simplicity, that $0 < A < 1$, so that we expect the predictor $x_t^{(0)}$ to tend to zero with increasing t. We must then have

$$x_t^{(0)} = \alpha^t x_0$$

where α is the smaller root of

$$(1 + A^2 + \theta NR) - A(\alpha + \alpha^{-1}) = 0 \tag{6.19}$$

If θ is positive then $0 < \alpha < A$, and the predictor is optimistic in that it tends to zero faster (and so incurs a smaller cost $\sum R x_t^2$) than does the risk-neutral predictor. If θ is negative then $\alpha > A$ and the predictor is pessimistic.

One can regard the prediction as having broken down altogether if θ is so negative that equation (6.19) has no root smaller than unity, when the extremized stress is infinite. This occurs when

$$\theta \leq \bar{\theta} = -\frac{(1-A)^2}{NR}$$

In fact, this is the value at which the saddle-point assumption of Theorem 6.2.1 fails. Its crossing marks a point at which the optimizer has become so pessimistic that he essentially gives up; see Section 6.4.

(3) Consider a risk-sensitive version of a pure estimation problem. Assume that x and y are jointly normal random vectors with zero mean and covariance matrix

$$\operatorname{cov}\begin{bmatrix} x \\ y \end{bmatrix} = \begin{bmatrix} I_{xx} & I_{xy} \\ I_{yx} & I_{yy} \end{bmatrix}^{-1}$$

and that there is a cost function $x'Rx$. The variable y has been observed, and x is to be estimated. In a risk-sensitive formulation one then estimates x by minimizing

$$\theta \mathbb{S} = \theta x' R x + \begin{bmatrix} x \\ y \end{bmatrix}' \begin{bmatrix} I_{xx} & I_{xy} \\ I_{yx} & I_{yy} \end{bmatrix}^{-1} \begin{bmatrix} x \\ y \end{bmatrix}$$

so deriving an estimate

$$x = -(I_{xx} + \theta R)^{-1} I_{xy} y$$

This estimate approaches the minimal cost value $x = 0$ as one increases θ. The matrix I_{xx} is the information matrix of x conditional on y; the effect of risk sensitivity is to modify this to $I_{xx} + \theta R$. That is, one 'gains' information if θ is positive (in that one is happy to let cost pressures 'supply' the information that x is nearer to its cost-minimizing value than the data y would have suggested). One 'loses' information if θ is negative (in that fear of costs makes one less certain that x is as near to the cost-minimizing value as the data would have suggested). When θ becomes so negative that $I_{xx} + \theta R$ is singular, estimation fails altogether; see Section 6.4.

6.4 OPTIMISM, PESSIMISM, NEUROSIS AND EUPHORIA

Investigation of particular problems will yield definite conclusions, but one can see in a general way that the risk-seeking and risk-averse cases really do correspond to an optimistic or pessimistic attitude on the part of the optimizer. The estimates of unobservables are chosen to extremize $\mathbb{C} + \theta^{-1}\mathbb{D}$. If θ is positive they will then be chosen with a consideration to making $\mathbb{C}$ small as well as $\mathbb{D}$. That is, the optimizer behaves as if unobservables would take values to his advantage; surely an attitude which defines optimism. If θ is negative then estimates will be chosen with a consideration to making $\mathbb{C}$ large as well as $\mathbb{D}$ small. That is, the optimizer behaves as if unobservables would take values to his disadvantage: pessimism.

We shall see that, as θ becomes progressively more positive, control will become ever slacker (being optimistic), to the point that it may become unstable if the uncontrolled system is unstable. However, a degree of optimism may be desirable, for it is often asserted that the conventional risk-neutral treatment gives excessive weight to the occasional large deviation, relative to the run of typical, moderate deviations.

As θ becomes ever more negative, control will become ever tighter (being pessimistic). There is a limit to the distance one may go in this direction. For any system with a non-negative definite cost function there is a critical (and negative) value of θ at which optimization fails. This is exactly the value at which the stress $\mathbb{S}$, a quadratic function of system variables, fails to possess a saddle point. At this value, $\bar{\theta}$, say, the criterion $\gamma_\pi(\theta)$ is infinite for all π, and the optimization problem becomes meaningless for $\theta < \bar{\theta}$.

One can identify $\bar{\theta}$ as corresponding to the level of pessimism at which the fear of disaster first outweighs the reassurance of statistics; the optimizer becomes convinced of his inability to control. It is not too fanciful to characterize this point as marking the onset of neurotic breakdown of the optimizer.

Theorem 6.4.1 (i) *The optimizer is, respectively, optimistic and pessimistic in the risk-seeking ($\theta > 0$) and risk-averse ($\theta < 0$) cases, in that he behaves as if unobservables would take values to his advantage or disadvantage, respectively.*
(ii) *Let $\bar{\theta}$ be the infimal value of θ at which stress $\mathbb{S} = \mathbb{C} + \theta^{-1}\mathbb{D}$ possesses a saddle point. If the quadratic function $\mathbb{C}$ is positive definite then this value is finite, always negative, and also characterized by*

$$\inf_\pi \gamma_\pi(\theta) \begin{cases} < +\infty & (\theta > \bar{\theta}) \\ = +\infty & (\theta < \bar{\theta}) \end{cases} \tag{6.20}$$

This value marks a degree of pessimism at which the optimizer becomes convinced of his inability to control.

Proof. Assertion (i) is a rather qualitative one, sufficiently supported by our discussion above.

For assertion (ii), since $\mathbb{D} = \xi'V\xi$ is positive-definite in ξ, then $\mathbb{S}$ certainly has a finite minimum in ξ for $\theta > 0$, and certainly fails to have a finite maximum in ξ for sufficiently negative θ; for $\theta < \bar{\theta}$, say. This $\bar{\theta}$ is just the negative value of θ at which $\mathbb{S}$ loses its saddle point. Also, the RSCEP evaluation fails at this point, since

$$\mathbb{S}_0 \begin{cases} < +\infty & (\theta > \bar{\theta}) \\ = +\infty & (\theta < \bar{\theta}) \end{cases} \qquad (6.21)$$

Note, however, that even for values of θ less than $\bar{\theta}$ one may still have finite $\mathbb{S}_t$ for some $t > 0$. One could interpret this either by saying that $\mathbb{S}$ may well still have a saddle point in a restricted set of variables or that pessimism has less purchase if one has more information.

We have still to establish that $\bar{\theta}$ is also characterized by (6.20). Suppose $\mathbb{S}_{t+1}$ well defined; we then have

$$-G(W_t) \propto \min_{u_t} \int \exp\left(-\frac{1}{2}\mathbb{S}_{t+1}\right) dy_{t+1} \propto \exp\left[-\frac{1}{2}\theta\left(\min_{u_t} \max_{y_{t+1}} \mathbb{S}_{t+1}\right)\right] \quad (6.22)$$

The second and third expressions in (6.22) will be finite or infinite together. That is, both $G(W_t)$ and $\mathbb{S}_t$ will be finite if and only if

$$\max_{y_{t+1}} \mathbb{S}_{t+1}$$

is finite. Working back recursively, we see then that

$$\inf_\pi \gamma_\pi(\theta) = -(2/\theta)\log G(W_0)$$

and $\mathbb{S}_0$ are either both infinite or both finite, whence relations (6.20) follow. $\square$

We know from Section 3.10 that there can be cases in which the cost function $\mathbb{C}$ is not positive-definite, but has a mixed character. We shall see in Exercise 6.4.2 and Section 7.8 that in such cases it is possible that optimization breaks down when θ *exceeds* some *positive* value $\bar{\theta}$. These tend to be cases in which one is maximizing a utility, and the interpretation would be that for $\theta > \bar{\theta}$ the optimizer is optimistic to the point of euphoria, and sees infinite riches beckoning him.

6.4.1 Exercises and comments

(1) Direct examination of multi-stage examples is best left until the analysis of the next chapter is completed, but examination of a single-stage example reveals

6.4 OPTIMISM, PESSIMISM, NEUROSIS AND EUPHORIA

clearly the nature of the dependence of control on θ and the occurrence of breakdown.

Assume that state and control variables are both scalar, and that one wishes to minimize $Qu^2 + \Pi x_1^2$, where $x_1 = Ax + Bu + \varepsilon$, initial state x is known, and ε has variance N. We assume Q, Π and N all positive. The expression for the stress is then

$$\mathbb{S} = Qu^2 + \Pi(Ax + Bu + \varepsilon)^2 + (\theta N)^{-1}\varepsilon^2$$

Regarding this as a quadratic form in u and ε, we see that the second-order terms have matrix

$$\begin{bmatrix} Q + B^2\Pi & B\Pi \\ B\Pi & \Pi + (\theta N)^{-1} \end{bmatrix}$$

We require that this be positive definite if θ is positive (which it is) and to possess a saddle point (i.e. be positive definite in u and negative definite in ε) if θ is negative. We can always minimize out u (because $Q + B^2\Pi > 0$) and obtain a quadratic expression in ε. In this ε^2 has coefficient $\Pi Q(Q + B^2\Pi)^{-1} + (\theta N)^{-1}$. We require that this coefficient have the same sign as θ, which it does just so long as θ exceeds $\bar{\theta}$, where

$$\bar{\theta} = -\frac{1}{N}\left[\frac{B^2}{Q} + \frac{1}{\Pi}\right] \tag{6.23}$$

Expression (6.23) indeed determines the critical value of θ, at which breakdown occurs.

The actual optimizing value of u is

$$u = -\frac{AB\Pi x}{Q + B^2\Pi + \theta N \Pi Q}$$

This becomes numerically smaller as θ becomes large and positive, reflecting the relaxation of an optimistic controller. It tends numerically to infinity as θ decreases towards $\bar{\theta}$; the pessimistic controller must take stronger action if he believes that Fate conspires against him.

(2) Consider Exercise (1) in the case $\Pi < 0$, so that the optimizer wishes x_1 to be large. If $\Pi < -Q/B^2$ then this desire outweighs control costs; he will take u as large as possible. If $-Q/B^2 < \Pi < 0$ then u will be chosen finite, but optimization will fail if θ *exceeds* the value $\bar{\theta}$ of (6.23), now positive. In this case the optimizer takes finite actions himself, but is so optimistic that he believes ε will take values infinitely to his advantage.

(3) The example of Exercise 6.3.1(2) is an uncontrolled one, so our requirement that the quadratic form (6.18) in the unobservables $x_1, x_2, \ldots$ should possess a saddle point for negative θ reduces to the requirement that this form should be negative definite. This condition fails, as we observed, unless θ exceeds the value $\bar{\theta}$ there evaluated.

6.5 OTHER INTERPRETATIONS OF THE LEQG CRITERION

Analyses based upon the assumption of Gaussian statistics generally still have some kind of 'second-order' interpretation, even if the Gaussian assumption is dropped. For example, the maximum likelihood estimate is still interpretable as a linear least-square estimate (Theorem 2.4.1) and the risk-neutral certainty-equivalence principle remains valid under the assumption of linear control rules (Theorem 2.5.3).

We have seen that LEQG-optimality implies the criterion that controls and unobservables should extremize stress:

$$\mathbb{S} = \mathbb{C} + \theta^{-1}\mathbb{D} \tag{6.24}$$

If we now make no Gaussian assumptions we could take this *conclusion* as a *starting point*. Essentially, one is choosing to minimize $\mathbb{C}$ subject to a constraint on the magnitude of $\mathbb{D}$, the parameter θ^{-1} playing the role of the Lagrange multiplier associated with this constraint.

The case of positive θ corresponds to the minimization of $\mathbb{C}$ (with respect to controls and unobservables) subject to a constraint

$$\mathbb{D} \leq d \tag{6.25}$$

where d is some prescribed value. One is thus willing to let unobservables 'help' control, an increasing d corresponding to an increasing latitude.

The case of negative θ corresponds to the situation where one minimizes/maximizes $\mathbb{C}$ with respect to controls/unobservables subject to condition (6.25). This is a min-max approach: one asks what the best controls may be, on the supposition that unobservables take the least favourable values (to the optimizer) within the class determined by (6.25). As one increases d, one increases this class, and so increases the contrariness of the unobservables which the controls must countervail.

Lastly, it turns out, remarkably, that control optimization in the H_∞ norm, which has awoken so much interest in recent years, is nothing but the LEQG-optimal control at the breakdown value, $\theta = \bar{\theta}$. Suppose that $\mathbb{C}$ is positive definite, so that $\bar{\theta}$ is negative. In adopting the LEQG criterion with negative θ one can be said to be looking for controls which are most successful in countervailing disturbances which are least favourable. As θ decreases down to $\bar{\theta}$ a pattern of input disturbances emerges which is critical. The LEQG-optimal controls at this point are just the H_∞-optimal controls. This connection was quite unsuspected until demonstrated by Glover and Doyle (1988); we shall develop it in Chapter 17.

There is a point to be noted. We have seen already (e.g. Exercise 6.4.1) that the control may become infinite as θ approaches the breakdown value. However, suppose, as is usual in the H_∞ literature, that one is working in the time-invariant case. For the constrained problem one is then extremizing the *average* instantaneous cost subject to a bound d on *average* discrepancy. Suppose also that the problem is homogeneous, in that one is regulating to zero, with zero deterministic disturbance α. Then, so long as d is positive, a change in its value is irrelevant, since this merely induces a scale change in all variables. This is the reason why, of all the values that θ might take, it is the breakdown value $\bar{\theta}$ which is significant, and why, in the time-invariant case, the optimal control at this value can be proper.

LEQG-optimal controls for larger values of θ relate, in the infinite-horizon limit, to the *minimal-entropy* controls; see Chapter 17.

6.6 NOTES ON THE LITERATURE

The first investigation of the LEQG criterion, and so of risk sensitivity in the LQG context, seems to have been that of Jacobson (1973, 1977). He and other authors worked on the state-structured case, but could treat only cases of perfect state observation, and some other special cases (see the literature notes for Chapters 7 and 8). It was the risk-sensitive certainty-equivalence principle which was lacking. This was supplied by Whittle, first for the state-structured case (1981) and then for the general case of Theorem 6.2.1 (1986a). The relation of the LEQG criterion to H_∞ and entropy criteria was first demonstrated by Glover and Doyle (1988).

CHAPTER 7

The Markov case: future stress and control

7.1 HYPOTHESES

Throughout this chapter and the next we shall assume the model of imperfectly observed state structure set out in hypotheses (M1–4)' of Section 4.1, and the exponential-of-quadratic performance criterion $\gamma_\pi(\theta)$ defined in equation (6.1).

The cost function $\mathbb{C}$ thus has the form

$$\mathbb{C} = \sum_{t=0}^{h-1} c_t + \mathbb{C}_h \tag{7.1}$$

where

$$c_t = \left[\begin{bmatrix} x \\ u \end{bmatrix}' \begin{bmatrix} R & S' \\ S & Q \end{bmatrix} \begin{bmatrix} x \\ u \end{bmatrix} \right]_t \tag{7.2}$$

$$\mathbb{C}_h = [x' \Pi x]_h \tag{7.3}$$

Note that we have abbreviated $c(x_t, u_t)$ to c_t; this is convenient. The discrepancy function $\mathbb{D}$ likewise has the form

$$\mathbb{D} = \mathbb{D}_0 + \sum_{t=1}^{h} d_t \tag{7.4}$$

Here the component

$$\mathbb{D}_0 = [(x - \hat{x}) V^{-1} (x - \hat{x})]_0 \tag{7.5}$$

arises from the prior distribution of x_0, and

$$d_t = \left[\begin{bmatrix} \varepsilon \\ \eta \end{bmatrix}' \begin{bmatrix} N & L \\ L' & M \end{bmatrix}^{-1} \begin{bmatrix} \varepsilon \\ \eta \end{bmatrix} \right]_t \tag{7.6}$$

from the distribution of (x_t, y_t) conditional on (X_{t-1}, Y_{t-1}). In expression (7.6) it is understood that ε and η are to be expressed in terms of x, y and u by appeal to the plant and observation relations

$$x_t = A x_{t-1} + B u_{t-1} + \alpha_t + \varepsilon_t \tag{7.7}$$

$$y_t = C x_{t-1} + \eta_t \tag{7.8}$$

THE MARKOV CASE: FUTURE STRESS AND CONTROL

The form of the cost function (7.1)–(7.3) implies that the problem is one of regulation to $x = 0$, $u = 0$. We know (Section 3.3) that the problem of regulating to an assigned path can be reduced to this case by redefinition of x, u and α.

Under the above assumptions the total stress is represented as a sum of terms over time:

$$\mathbb{S} = \mathbb{C} + \theta^{-1}\mathbb{D} = \theta^{-1}\left(\mathbb{D}_0 + \sum_1^h d_t\right) + \sum_0^{h-1} c_t + \mathbb{C}_h \tag{7.9}$$

One can then meaningfully speak of *past stress* and *future stress*, and thereby decompose the problem.

7.2 PAST AND FUTURE STRESSES

Define the extremized past stress

$$P_t(x_t, W_t) = \underset{x_0,\ldots,x_{t-1}}{\text{ext}} \left(\theta^{-1}\mathbb{D}_0 + \theta^{-1}\sum_1^t d_\tau + \sum_0^{t-1} c_\tau\right) \tag{7.10}$$

and the extremized future stress

$$F_t(x_t) = \min_{u_t,\ldots,u_{h-1}} \underset{\substack{x_{t+1},\ldots,x_h \\ y_{t+1},\ldots,y_h}}{\text{ext}} \left(\theta^{-1}\sum_{t+1}^h d_\tau + \sum_t^{h-1} c_\tau + \mathbb{C}_h\right) \tag{7.11}$$

Here we have split up the total stress (7.9) into components which can be allocated to past and future at time t. Note that d_t has been allocated to the past and c_t to the future, because of our convention that at time t the observation y_t is available but the decision u_t has not yet been taken. We have minimized out decisions currently unmade and θ-extremized out all current unobservables *with the exception* of x_t, the current state variable. The current state variable is *pivotal*, in that it is the only variable to be extremized which appears in both past and future stresses. By holding it as a free variable for the moment we can decouple the extremizations of past and future stress.

It is convenient, although slightly abusive of nomenclature, to refer to the functions P_t and F_t simply as *past stress* and *future stress*, respectively, although they are in fact partially extremized versions of these quantities.

We shall use $\check{x}_t$ to denote the MSE $x_t^{(t)}$, i.e. the estimate extremizing *total* stress, and $\hat{x}_t$ to denote the estimate extremizing *past* stress at time t. These two estimates coincide in the risk-neutral case.

The past stress P_t evidently obeys the forward recursion

$$P_t(x_t, W_t) = \underset{x_{t-1}}{\text{ext}} \left[\theta^{-1}d_t + c_{t-1} + P_{t-1}(x_{t-1}, W_{t-1})\right] \tag{7.12}$$

with initial condition

$$P_0(x_0, W_0) = \theta^{-1}\mathbb{D}_0 \tag{7.13}$$

7.2 PAST AND FUTURE STRESSES

The future stress F_t correspondingly obeys the backward recursion

$$F_t(x_t) = \min_{u_t} \underset{x_{t+1}, y_{t+1}}{\text{ext}} [\theta^{-1} d_{t+1} + c_t + F_{t+1}(x_{t+1})] \tag{7.14}$$

with terminal condition

$$F_h(x_h) = \mathbb{C}_h \tag{7.15}$$

We shall see that these reduce to interesting variants of forms familiar from the risk-neutral case.

Theorem 7.2.1. (i) *The separation principle. The evaluation of extremal past and future stress can be decoupled if these evaluations are made conditional on the value of current state x_t. These partially extremized stresses, $P_t(x_t, W_t)$ and $F_t(x_t)$, relate to estimation and control, respectively, in that evaluation of P_t summarizes the effect of past observations, and that the evaluation of F_t will also imply the evaluation of the control $u_t(x_t)$ which would be optimal if x_t were known.*

These calculations are recoupled by the extremization of $P_t(x_t, W_t) + F_t(x_t)$ with respect to x_t to yield the minimal stress estimate $\check{x}_t$.

(ii) *The certainty-equivalence principle. The optimal value of u_t is $u_t(\check{x}_t)$.*

Proof. The various assertions of (i) are evident from the previous discussion. As for assertion (ii), $u_t(\check{x}_t)$ is just the MSE $u_t^{(t)}$ of u_t, and so the optimal value of u_t, by Theorem 6.2.1. As an evaluation it is notable only in that, in the joint extremization of stress with respect to controls and unobservables, x_t is extremized last. □

Assertion (i) indeed describes how the separation principle must now be understood; as a provisional decoupling of past/estimation considerations from future/control considerations by specification of the pivotal variable: current state. The interaction of estimation and control comes in the recoupling.

This recoupling yields assertion (ii), which looks much more like the conventional statement of certainty equivalence than does the general Theorem 6.2.1. In the risk-sensitive as well as the risk-neutral state-structured case we can interpret $u_t(x_t)$ as the optimal control if x_t is known, and $u_t(x_t^{(t)})$ as the optimal control if x_t is unobservable. The difference in the two cases is that in the risk-sensitive case the estimate $x_t^{(t)}$ depends upon costs along the whole path as well as on past observations.

We shall see, in this chapter and the next, that the recursion (7.14) for future stress is very close to the recursion (3.16) for value functions and optimal control with perfect state observation, and that the recursion (7.12) for past stress is very close to the recursions behind Theorem 4.2.1, which update the posterior distribution of state.

7.2.1 Exercise and comments

(1) Assume that $y_t = x_{t-1}$, so that x_{t-1} is perfectly observed at time t. Show that then
$$P_t(x_t, W_t) = \theta^{-1}\varepsilon_t' N^{-1}\varepsilon_t + \cdots$$
where ε is to be expressed in terms of x and u from the plant equation and $+\cdots$ indicates terms independent of x_t.

7.3 SOLUTION OF THE F-RECURSION

We shall speak of recursions (7.12) and (7.14) as the P- and F-recursions, respectively. As indicated at the end of the previous section, these are the analogues of the recursions used to determine optimal estimate and optimal control, respectively, in the risk-neutral case. We can continue so to think of them, provided we remember that both estimate and control are modified when the two calculations are recoupled.

The F-recursion must be solved in any case, and we shall begin with this. However, note that the F-recursion is all that need be solved if current state is indeed observable.

We shall be speaking of future values x_τ, y_τ ($\tau > t$), but these are stress-extremizing values for a given value of x_t. Thus the quantity we shall denote by x_τ, for example, is neither the actual value x_τ nor the MSE $x_\tau^{(t)}$, but the MSE conditional on prescribed x_t.

The extremization with respect to y_τ in (7.11) reduces d_τ to
$$\min_{\eta_\tau} d_\tau = (\varepsilon' N^{-1}\varepsilon)_\tau \qquad (t < \tau \le h) \tag{7.16}$$
which implies a prediction of y_τ (see Exercise 7.3.1(1)). The F-recursion thus reduces to the explicit form
$$F_t(x_t) = \min_{u_t} \text{ext}_{x_{t+1}} [c(x_t, u_t) + \theta^{-1}(\varepsilon' N^{-1}\varepsilon)_{t+1} + F_{t+1}(x_{t+1})] \tag{7.17}$$
with x_{t+1} and ε_{t+1} related by the plant equation (7.7) and $c(x, u)$ having the familiar quadratic form (7.2). The fact that we are extremizing out both future u- and x-values is an indication that the exercise is one in the combined optimization of control and prediction.

Appealing to the plant equation, we can eliminate x_{t+1} to write (7.17) as
$$F_t(x_t) = \min_{u_t} \text{ext}_{\varepsilon_{t+1}} [c(x_t, u_t) + \theta^{-1}(\varepsilon' N^{-1}\varepsilon)_{t+1} + F_{t+1}(Ax_t + Bu_t + \alpha_{t+1} + \varepsilon_{t+1})] \tag{7.18}$$

The form of (7.18) makes clear that ε_{t+1} acts effectively as a supplementary control: one that is to chosen to help u_t if θ is positive and to frustrate u_t if θ is negative.

Note that the ε-extremization must be applied *before* the u-minimization. This

7.3 SOLUTION OF THE F-RECURSION

is because in the full optimality equation (such as (6.11)) one would take a conditional expectation over the future value of x_{t+1} *before* optimizing the current control u_t, and this expectation corresponds to the ε-extremization.

Recursion (7.18) can be regarded as obtained by the application of two operators;

$$F_t = \mathscr{L}_t \tilde{\mathscr{L}} F_{t+1} \tag{7.19}$$

Here $\mathscr{L}_t$ is just the operator of the risk-neutral (and noise-free) case:

$$\mathscr{L}_t \phi(x) = \min_u \left[c(x,u) + \phi(Ax + Bu + \alpha_{t+1}) \right]$$

and $\tilde{\mathscr{L}}$ is the operator with effect

$$\tilde{\mathscr{L}} \phi(x) = \underset{\varepsilon}{\text{ext}} \left[\theta^{-1}(\varepsilon' N^{-1} \varepsilon) + \phi(x + \varepsilon) \right] \tag{7.20}$$

To take the example which will be relevant, assume that $\phi(x)$ is a quadratic form, $x' \Pi x$. Then we find that expression (7.20) has the evaluation $x' \tilde{\Pi} x$, where $\tilde{\Pi} = \tilde{f} \Pi$ is a transformation which can be variously expressed:

$$\tilde{\Pi} = \tilde{f} \Pi = (\Pi^{-1} + \theta N)^{-1} = \Pi(I + \theta N \Pi)^{-1} = (I + \theta \Pi N)^{-1} \Pi \tag{7.21}$$

This evaluation is not valid unconditionally. In the following we take for granted that $0 \leq N < \infty$.

Theorem 7.3.1. *Assume that $0 \leq \Pi < \infty$. Then the evaluation $\tilde{\mathscr{L}}(x' \Pi x) = x' \tilde{\Pi} x$ is valid iff $0 \leq \tilde{\Pi} < \infty$; equivalently, iff*

$$\Pi^{-1} + \theta N > 0 \tag{7.22}$$

Proof. The form $\theta^{-1}(\varepsilon' N^{-1} \varepsilon) + (x + \varepsilon)' \Pi (x + \varepsilon)$ is extremal for a finite value of ε if

$$\begin{aligned} \Pi + (\theta N)^{-1} &> 0 \quad (\theta > 0) \\ \Pi + (\theta N)^{-1} &< 0 \quad (\theta < 0) \end{aligned} \tag{7.23}$$

There exists a co-ordinate transformation which simultaneously diagonalizes Π and N; we may therefore assume that we are in the scalar case. Condition (7.23) is then equivalent to

$$\theta \Pi + N^{-1} > 0 \tag{7.24}$$

this being trivially valid if $\theta = 0$. In fact, if either $\theta \geq 0$ or $N = 0$ then relations (7.22) and (7.24) are trivially valid, under the hypothesis on Π. If $N > 0$ and $\theta < 0$ then (7.22) and (7.24) are equivalent. □

The value of ε which achieves the extremum in (7.20) is

$$\varepsilon = -\tilde{\Pi}^{-1} \Pi x \tag{7.25}$$

Collecting these observations we have:

98   THE MARKOV CASE: FUTURE STRESS AND CONTROL

Theorem 7.3.2. *Consider the F-recursion* (7.18) *in the homogeneous case* $\alpha = 0$. *It has the solution*

$$F_t(x) = x' \Pi_t x \tag{7.26}$$

where Π_t satisfies the modified Riccati equation

$$\Pi_t = f\tilde{f} \Pi_{t+1} \tag{7.27}$$

with prescribed terminal value Π_h ($0 \leq \Pi_h < \infty$). Here f is the operator of the risk-neutral Riccati equation, defined in (3.14), *and $\tilde{f}$ has the action* (7.21), *defined if*

$$\Pi_{t+1}^{-1} + \theta N > 0 \tag{7.28}$$

The matrix Π_t and the matrix K_t of the optimal control $u_t = K_t x_t$ will then be determined in terms of Π_{t+1} as in (3.13)–(3.15) (*or* (3.39) *and* (3.40)), *except that $\tilde{\Pi}_{t+1} = (\Pi_{t+1}^{-1} + \theta N)^{-1}$ replaces Π_{t+1}.*

Therefore results differ from the risk-neutral case only by the insertion of the extra transformation $\Pi_{t+1} \to \tilde{f}\Pi_{t+1} = \tilde{\Pi}_{t+1}$ in the recursions. The effect of this is particularly transparent if we use the recursions in the alternative forms of Section 3.5.

7.3.1 Exercises and comments

(1) The minimization (7.16) implies a prediction relation $\eta_\tau = L'N^{-1}\varepsilon_\tau$ or

$$y_\tau = Cx_\tau + L'N^{-1}(x_\tau - Ax_{\tau-1} - Bu_{\tau-1} - \alpha_\tau) \qquad (\tau > t)$$

where all quantities are minimal-stress predictions assuming knowledge of x_t.
(2) Suppose the condition on Π in Theorem 7.3.1 replaced by $-\infty < \Pi \leq 0$. Show that the condition for validity of the transformation $\tilde{f}$ then becomes $\Pi^{-1} + \theta N < 0$, a condition having force only for θ positive. What is the condition for general symmetric finite Π?

7.4 ALTERNATIVE FORMS: THE MODIFIED CONTROL-POWER MATRIX

By calculating the transformation $f\tilde{f}\Pi$ with f given the alternative form stated in (3.40) we deduce the risk-sensitive version of these alternative relations.

Theorem 7.4.1. *In the risk-sensitive case (with $\alpha = 0$) the dependences of K_t and Π_t on Π_{t+1} have exactly the forms* (3.39) *and* (3.40), *except that the control-power matrix J is modified to*

$$J = BQ^{-1}B' + \theta N \tag{7.29}$$

7.5 THE CONSTANT TERM IN FUTURE COSTS

The extremizing value of ε_{t+1} *is*

$$\varepsilon_{t+1} = -\theta N(J + \Pi_{t+1}^{-1})^{-1}A^*x \tag{7.30}$$

Recall that $A^* = A - BQ^{-1}S$, the transformed value of A under the normalization of S to zero. The calculations are immediate, and the effects of risk sensitivity evident. As we see from (7.29), the effect is to modify the *control-power matrix* $BQ^{-1}B'$ by the addition of θ times the *noise-power matrix* N. Therefore noise power has the effect of either supplementing or reducing control power, according as to whether θ is negative or positive. This constitutes a nice summary of the effects we have already observed.

Let us write the risk-neutral form of the control-power matrix as

$$J_0 = BQ^{-1}B' \tag{7.31}$$

We then have an *actual* gain matrix

$$\Gamma_t = A + BK_t = A^* - J_0(J + \Pi_{t+1}^{-1})^{-1}A^* \tag{7.32}$$

However, the predicted value of x_{t+1} will be given, not just by $Ax_t + Bu_t = \Gamma_t x_t$ but by $Ax_t + Bu_t + \varepsilon_{t+1}$, where ε_{t+1} will have the evaluation (7.30). We thus have a formal or *predictive* gain matrix for the MS-predicted course of the process

$$\tilde{\Gamma}_t = A^* - J(J + \Pi_{t+1}^{-1})^{-1}A^* \tag{7.33}$$

This predictive gain matrix takes account of the subsidiary 'control' exerted by ε; it is this matrix which we should expect to be a stability matrix in the infinite-horizon limit. We can write it also in the forms

$$\tilde{\Gamma}_t = (I + \theta N \Pi_{t+1})^{-1}\Gamma_t = \Pi_{t+1}^{-1}\tilde{\Pi}_{t+1}\Gamma_t \tag{7.34}$$

Note that the use of these alternative forms does imply a permutation of the various operations which will be valid only if the saddle-point condition is fulfilled. We saw from Theorem 7.3.1 that in the present case this condition amounts to

$$\Pi_{t+1}^{-1} + \theta N > 0 \tag{7.35}$$

for all relevant t (if $\Pi_{t+1} \geq 0$). We shall term this the *risk-resistance condition*, because it implies that the cost function $x'\Pi_{t+1}x$ is not so rapidly increasing in x that it attracts an infinite value of disturbance ε_{t+1} from a malign Nature.

7.5 THE CONSTANT TERM IN FUTURE COSTS

Let us define the future cost at time t as

$$\mathbb{C}_t = \sum_{\tau=t}^{h-1} c_\tau + \mathbb{C}_h \tag{7.36}$$

and define the risk-sensitive value function

$$\bar{F}_t(W_t) = \inf_\pi \left[-(2/\theta)\log\left(E_\pi\left[\exp\left(-\frac{1}{2}\theta \mathbb{C}_t\right) \bigg| W_t \right]\right) \right] \quad (7.37)$$

For the moment we are assuming full state information: that knowledge of W_t implies knowledge of x_t. We assume also the homogeneous case: $\alpha = 0$.

The future stress F_t calculated in Section 7.3 is then in fact the x_t-dependent part of $\bar{F}_t$. There will also be a constant term, reflecting the component of future cost due to future noise. We have tended to neglect such terms, as not affecting policy, but there will be purposes for which we should evaluate them.

Theorem 7.5.1. *$\bar{F}_t$ is a function $\bar{F}_t(x_t)$ of x_t and t alone, with evaluation*

$$\bar{F}_t(x_t) = F_t(x_t) + \gamma_t \quad (7.38)$$

where

$$\gamma_t - \gamma_{t+1} = \theta^{-1} \log|I + \theta N \Pi_{t+1}| \quad (7.39)$$

Proof. The (x_t, t) dependence of $\bar{F}_t$ follows from the usual inductive arguments. It will obey the recursion

$$\exp\left[-\frac{1}{2}\theta \bar{F}_t(x)\right] = \underset{u}{\text{ext}}\, (2\pi)^{-n/2}|N|^{-1/2} \int \exp\left[-\frac{1}{2}\varepsilon' N^{-1}\varepsilon \right.$$
$$\left. -\frac{1}{2}\theta[c(x,u) + \bar{F}_t(Ax + Bu + \varepsilon)]\right] d\varepsilon \quad (7.40)$$

From (7.40) we verify inductively that $\bar{F}_t$ has the form

$$\bar{F}_t(x) = x' \Pi_t x + \gamma_t \quad (7.41)$$

where Π satisfies the recursion (7.27) and γ satisfies (7.39). □

Note that in the limit of zero θ (7.39) becomes

$$\gamma_t - \gamma_{t+1} = \text{tr}(N \Pi_{t+1})$$

consistently with Exercise 3.2.1(5).

7.6 THE DISTURBED (NON-HOMOGENEOUS) CASE

If the known disturbance α_t is retained in the plant equation (7.1) then we will expect the future stress F_t to have the non-homogeneous quadratic form

$$F_t(x) = x' \Pi_t x - 2\sigma' x + \cdots \quad (7.42)$$

where $+ \cdots$ indicates terms independent of x. To develop recursions we note first

Lemma 7.6.1

$$\mathscr{L}(x'\Pi x - 2\sigma'x) = x'\tilde{\Pi}x - 2\tilde{\sigma}'x + \cdots \tag{7.43}$$

where $\tilde{\Pi}$ has the evaluation $(\Pi^{-1} + \theta N)^{-1}$ of (7.21) and

$$\tilde{\sigma} = (I + \theta\Pi N)^{-1}\sigma = \tilde{\Pi}\Pi^{-1}\sigma \tag{7.44}$$

Verification is immediate.

Theorem 7.6.2. *In the case of a deterministically disturbed plant equation the future stress $F_t(x)$ has the form (7.42) with Π_t obeying the modified Riccati equation indicated in Theorems 7.3.1 and 7.4.1 and σ_t obeying the recursion*

$$\sigma_t = \Gamma_t'(\tilde{\sigma}_{t+1} - \tilde{\Pi}_{t+1}\alpha_{t+1}) = \tilde{\Gamma}_t'(\sigma_{t+1} - \Pi_{t+1}\alpha_{t+1}) \tag{7.45}$$

Here Γ_t and $\tilde{\Gamma}_t$ are the actual and predictive gain matrices defined in (7.32) and (7.33).

Verification is direct. We see that recursion (7.45) differs from the corresponding risk-neutral recursion (3.25) only by the substitution of $\tilde{\Gamma}_t$ for the risk-neutral evaluation of Γ_t. This indeed is what it must be. We have replaced optimization with respect to a single control u by optimization with respect to a double control (u, ε).

The same argument leads to the evaluation of the optimal control

$$u_t = K_t x_t + (Q + B'\tilde{\Pi}_{t+1}B)^{-1}B'(\tilde{\sigma}_{t+1} - \tilde{\Pi}_{t+1}\alpha_{t+1}) \tag{7.46}$$

with the feedback/feedforward expression

$$u_t = K_t x_t - (Q + B'\tilde{\Pi}_{t+1}B)^{-1}B'(I + \theta\Pi_{t+1}N)^{-1} \sum_{j=0}^{h-t-1} \tilde{\Gamma}'_{t+1}\tilde{\Gamma}'_{t+2}\cdots$$

$$\tilde{\Gamma}'_{t+j}\Pi_{t+j+1}\alpha_{t+j+1} \tag{7.47}$$

7.6.1 Exercise and comments

(1) We could express F_t as $(x - x_t^*)'\Pi_t(x - x_t^*) + \cdots$ rather than as (7.42) We would interpret x_t^* as the ideal value of x_t as far as future costs are concerned, because it minimizes $F_t(x)$. Recursion (7.45) then becomes

$$x_t^* = \Pi_t^{-1}\tilde{\Gamma}_t'\Pi_{t+1}(x_{t+1}^* - \alpha_{t+1})$$

7.7 THE CONTINUOUS-TIME ANALOGUE

We deduce the continuous-time analogues by the same limiting procedure as in Section 3.9. Then F will have the same quadratic form:

$$F(x) = x'\Pi x - 2\sigma'x + \cdots \tag{7.48}$$

where F, Π and σ all have a suppressed time dependence. The limit form of either of the Riccati equations for Π is

$$\dot{\Pi} + R + A'\Pi + \Pi A - (S' + \Pi B)Q^{-1}(S + B'\Pi) - \theta N = 0 \qquad (7.49)$$

This reduces to

$$\dot{\Pi} + R + A'\Pi + \Pi A - \Pi J \Pi = 0 \qquad (7.50)$$

if S has been normalized to zero. Here, as in Section 7.4,

$$J = J_0 + \theta N = BQ^{-1}B' + \theta N \qquad (7.51)$$

is the 'risk-tempered' control-power matrix.

The distinction between Π and $\tilde{\Pi}$ vanishes in the continuous limit; similarly, the distinction between σ and $\tilde{\sigma}$. In particular, the risk-resistance condition (7.28) is now no condition at all. In effect, for operations over a short time the 'controls' u and ε exert their effects almost simultaneously, and their 'optimizations' do not interact. Furthermore, it would be infinitely costly to choose an infinite value of ε. Therefore the only effect of risk sensitivity is to modify the basic Riccati equation (7.50).

The optimal control is then again (c.f. equation (3.86))

$$u = Kx + Q^{-1}B'\sigma \qquad (7.52)$$

where

$$K = -Q^{-1}(S + B'\Pi) \qquad (7.53)$$

and

$$\dot{\sigma} + \tilde{\Gamma}'\sigma - \Pi\alpha = 0 \qquad (7.54)$$

Here $\tilde{\Gamma}$ is the predictive gain matrix

$$\tilde{\Gamma} = A - BQ^{-1}S - \Pi J \qquad (7.55)$$

If we define the future stress criterion $\bar{F}$ by (7.37) then we have again that

$$\bar{F}(x) = x'\Pi x - 2\sigma'x + \gamma \qquad (7.56)$$

where the absolute cost term γ can be verified to obey

$$\dot{\gamma} + \text{tr}(N\Pi) - \sigma'\alpha - \alpha'\sigma - \sigma'J\sigma = 0 \qquad (7.57)$$

7.8 SOME EXAMPLES

Even the scalar case is interesting; we analyse this and its infinite-horizon behaviour in Section 7.9. In this section we consider some continuous-time examples.

(i) *The inertialess missile.* Consider the scalar case with plant equation $\dot{x} = u + \varepsilon$, instantaneous cost rate Qu^2 and terminal cost x^2. The aim is then to terminate at $x = 0$, and the only cost incurred along the path is the integral of the control cost Qu^2. One could regard x as the transverse deviation of a missile from a desired flight path of known duration. The fact that x follows first-order dynamics indicates that u has its immediate effect on velocity rather than on acceleration, and so is overcoming 'viscosity' rather than 'inertia'.

We have $F = \Pi x^2$, and can take Π as a function $\Pi(s)$ of time-to-go s, with $\Pi(0) = 1$. Equation (7.50) becomes simply

$$\mathring{\Pi} = -J\Pi^2 \tag{7.58}$$

where $\mathring{\Pi} = d\Pi/ds$ and

$$J = Q^{-1} + \theta N \tag{7.59}$$

This has solution

$$\Pi(s) = (1 + Js)^{-1} \tag{7.60}$$

and the optimal control is

$$u = -Q^{-1}(1 + Js)^{-1}x \tag{7.61}$$

The effect of varying θ upon costs and controls is evident from equations (7.59)–(7.61). We shall have finiteness of costs and controls for all s only if $J > 0$, i.e.

$$\theta > \bar{\theta} = -(NQ)^{-1} \tag{7.62}$$

If (7.62) does not hold then we reach a 'disaster point' as soon as $s \geq \bar{s} = -J^{-1} = -(Q^{-1} + \theta N)^{-1}$. Effectively, if $s \geq \bar{s}$ then there is enough time left before termination that the contrariness of Nature is considered able to ruin the programme.

(ii) *The inertial missile.* Suppose that x has components x_1 and x_2, representing deviation and rate of change of deviation from path. We assume the plant equation

$$\dot{x}_1 = x_2$$

$$\dot{x}_2 = u + \varepsilon$$

terminal cost x_1^2 and instantaneous cost rate Qu^2. This formulation then differs from that of (i) in that both control u and noise ε act on acceleration; they are forces imposed on a body with inertia. As is, in general, true for LQ problems in which the instantaneous cost is independent of state, the future stress F and the optimal control depend upon current state only through the 'predicted terminal miss-distance' $(x_1 + x_2 s)$: what the terminal miss-distance would be if no further forces were applied. Assume then that

$$F(x) = \Lambda(s)(x_1 + x_2 s)^2$$

with $\Lambda(0) = 1$. By either reducing the full Riccati equation (7.50) or by establishing an equation for Λ directly we find that

$$\mathring{\Lambda} = -Js^2\Lambda^2$$

with J again given by (7.59). We thus find that

$$\Lambda(s) = (1 + Js^3/3)^{-1}$$
$$u = -Q^{-1}(1 + Js^3/3)^{-1}s(x_1 + x_2 s)$$

Thus F and u are finite for all s as long as $J > 0$, i.e. if (7.62) holds. In other cases we have failure for $s \geq \bar{s} = (-3/J)^{1/3}$.

Note that neither problems (i) nor (ii) have proper infinite horizon limits even if $J > 0$. This is due to the fact that state deviations do not affect instantaneous costs. The apparent limit values $u = 0$ and $F = 0$ cannot really be regarded as informative. They merely state that, if infinite time remains, then no immediate action is called for, and action can be taken so gradually that no cost is incurred.

(ii) *The monopolist with sticky prices.* This is exactly the model discussed in Section 3.10, but with a disturbed plant equation

$$\dot{x} = a(p - x - u) + \varepsilon$$

and a risk-sensitive criterion. Assuming $\bar{F}(x) = \Pi x^2 - 2\sigma x + \gamma$ (which differs from future stress $F(x)$ only by terms independent of x) we find by appeal to the formulae of Section 7.7 that

$$\mathring{\Pi} = -[(a^2 + \theta N)\Pi^2 + 4a\Pi + 1] \tag{7.63}$$

$$\mathring{\sigma} = -[a(\Pi p + \sigma) + \theta N \sigma \Pi + (a\sigma + c)(1 + a\Pi)]$$

$$\mathring{\gamma} = N\Pi - 2ap\sigma - \theta N\sigma^2 - (a\sigma + c)^2 \tag{7.64}$$

In the infinite horizon limit we expect that $\mathring{\Pi}$ and $\mathring{\sigma}$ tend to zero, and that $\mathring{\gamma}$ tends to a constant which represents the cost, in the risk-sensitive sense, incurred per unit time because of noise. The equilibrium solutions of equation (7.63) are

$$\Pi = \frac{-2a \pm \sqrt{(3a^2 - \theta N)}}{a^2 + \theta N} \tag{7.65}$$

As far as θ is concerned there seem to be two values which might prove critical: the value

$$\theta_1 = -a^2/N$$

at which the roots change sign, and the value

$$\theta_2 = 3a^2/N$$

at which the roots become coincident.

As in Section 3.10, let us write equation (7.63) as $\mathring{\Pi} = \chi(\Pi)$. For $\theta > \theta_2$ we have $\chi(\Pi) < 0$ for all Π. (See Figure 4.) There is no equilibrium point, and $\Pi \to -\infty$ with increasing s.

7.8 SOME EXAMPLES

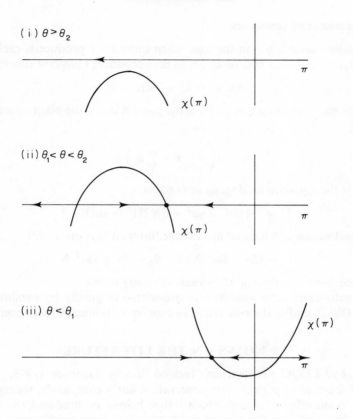

Figure 4. The solution of the equilibrium Riccati equation for the production problem, for different ranges of the risk-sensitivity parameter θ. Cases (a), (b) and (c) correspond to $\theta > \theta_2$, $\theta_1 < \theta < \theta_2$ and $\theta < \theta_1$ respectively

For $\theta_1 < \theta < \theta_2$ the situation is qualitatively as in the risk-neutral case of Section 3.10: both roots (7.65) are negative, and it is the upper one which is stable. For $\theta < \theta_1$ there is a positive root and a negative root, and it is the negative one which is stable.

Therefore in this case of negative definite forms (positive rewards) it is the value θ_2 which is critical; for $\theta > \theta_2$ the optimizer is euphoric, in that he believes he can achieve an infinite income (in that $\Pi \to -\infty$ and so $\gamma \to -\infty$, see (7.64)).

In the case $\theta_1 < \theta < \theta_2$ there is an equilibrium point, but the optimizer will still become euphoric if the terminal value of Π is less than the smaller root of (7.65); i.e. if terminal reward is too large. In the case $\theta < \theta_1$ there is an equilibrium point, but the optimizer becomes catastrophically pessimistic if the terminal value of Π exceeds the larger root of (7.65); i.e. if terminal penalty is too great.

7.8.1 Exercises and comments

(1) Consider example (iii) in the case when there are r producers, each having the same instantaneous cost function, in that producer i has cost function

$$c(x, u_i) = u_i^2 + 2u_i(c - x)$$

where u_i is his production rate and x is the price. Assume the plant equation then becomes

$$\dot{x} = a\left(p - x - \sum_i u_i\right) + \varepsilon$$

Show that the equation analogous to (7.63) is

$$\mathring{\Pi} = -[(3 - 2r)a^2 + \theta N]\Pi^2 - 4a\Pi - 1$$

The critical values of θ located in example (iii) then become

$$\theta_1 = (2r - 3)a^2/N, \qquad \theta_2 = (2r + 1)a^2/N$$

so that the 'euphoria threshold' increases linearly with r.

This multi-agent case was the one considered originally by Fershtman and Kamien (1987), and in the risk-sensitive case by Y. Willassen (pers. comm.).

7.9 NOTES ON THE LITERATURE

The Markov LEQG problem was tackled first by Jacobson (1973, 1977). He analysed the case of perfect state observation fairly completely, seeing process noise as a subsidiary control which either helped or hindered the intended control, depending on the sign of θ. He also demonstrated the occurrence of 'breakdown' at a critical negative value of θ.

The analysis of Section 7.2, which appeals to the RSCEP and formulates the separation principle as the possibility of decoupling the extremization of past and future stresses, is due to Whittle (1981). This article also first demonstrated the significance of the augmented control-power matrix (7.29).

References for the production problem with sticky prices are given in Exercise 7.8.1. See the literature notes of the next chapter for further comment on the case of partial observation.

CHAPTER 8

The Markov case: past stress and estimation

8.1 SOLUTION OF THE P-RECURSION: THE RISK-SENSITIVE KALMAN FILTER

We continue with the set of Markov LEQG assumptions listed in Section 7.1. If information on current state is imperfect then we have also to solve the P-recursion (7.12). We can write this in the form

$$\theta P_t(x_t, W_t) = \min_{x_{t-1}} \left[d_t + \theta c_{t-1} + \theta P_{t-1}(x_{t-1}, W_{t-1}) \right] \quad (8.1)$$

Here the cost and discrepancy components c and d are defined by equations (7.2) and (7.6) with the noise variables expressed in terms of x, y and u from the plant and observation relations.

Now, if we take the risk-neutral limit $\theta \to 0$ then θP_t will have a limit, D_t, say, satisfying the risk-neutral form of (8.1):

$$D_t(x_t, W_t) = \min_{x_{t-1}} \left[d_t + D_{t-1}(x_{t-1}, W_{t-1}) \right] \quad (8.2)$$

D_t is in fact

$$\min_{x_{t-1}} \left[-2 \log f(X_t, Y_t | U_{t-1}) \right]$$

to within a constant, and we have

$$D_t(x_t, W_t) = \left[(x - \hat{x})' V^{-1} (x - \hat{x}) \right]_t + \cdots \quad (8.3)$$

where $\hat{x}_t$ and V_t are the mean and covariance matrix of x_t conditional on W_t and $+ \cdots$ indicates terms independent of x_t. Recursion (8.2) implies recursions for $\hat{x}_t$ and V_t, which must be just the Kalman filter and the Riccati recursion in one form or another.

In the risk-sensitive case we can again establish that θP_t has the quadratic form (8.3), so that

$$P_t(x_t, W_t) = \theta^{-1} \left[(x - \hat{x})' V^{-1} (x - \hat{x}) \right]_t + \cdots \quad (8.4)$$

where $+ \cdots$ again indicates terms independent of x_t, irrelevant for updating purposes. The quantity $\hat{x}_t$ now denotes the estimate of x_t which θ-extremizes *past stress* at time t, to be distinguished from the total stress estimate $\check{x}_t$. V_t can no

108   THE MARKOV CASE: PAST STRESS AND ESTIMATION

longer be interpreted as a covariance matrix; one can say that it measures precision of $\hat{x}_t$ in that V_t^{-1} measures the curvature of past stress as x_t varies from $\hat{x}_t$.

Theorem 8.1.1. *The past stress has the quadratic form* (8.4) *with initial values $\hat{x}_0$ and V_0 identified as the prescribed mean and covariance of x_0 conditional on W_0. The values of $\hat{x}_t$ and V_t are determined from those of $\hat{x}_{t-1}$ and V_{t-1} by the Kalman filter* (4.8) *and the Riccati equation* (4.9) *of the risk-neutral case, with the modification that V_{t-1} is replaced by*

$$\tilde{V}_{t-1} = (V_{t-1}^{-1} + \theta R)^{-1} \tag{8.5}$$

and $\hat{x}_{t-1}$ by

$$\tilde{x}_{t-1} = (V_{t-1}^{-1} + \theta R)^{-1}(V_{t-1}^{-1}\hat{x}_{t-1} - \theta S' u_{t-1}) \tag{8.6}$$

For validity of these calculations it is necessary that

$$V_{t-1}^{-1} + \theta R > 0 \tag{8.7}$$

at all values of t at which information is updated.

That is, as for the case of the F-recursion, the recursions of the risk-sensitive case are just those of the risk-neutral case, except that another operation is inserted between the recursive steps.

Proof. Certainly the form (8.4) holds at $t = 0$ (c.f. (7.8)). Since all terms are quadratic in recursion (8.1) we see inductively that the form (8.4) holds for all t.

If we substitute expressions (7.2) and (8.4) for c_{t-1} and P_{t-1}, respectively, we find that

$$\theta(c_{t-1} + P_{t-1}) = [(x - \tilde{x})'\tilde{V}^{-1}(x - \tilde{x})]_{t-1} + \cdots \tag{8.8}$$

Here $\tilde{x}_{t-1}$ and $\tilde{V}_{t-1}$ have the evaluations (8.5) and (8.6) and $+\cdots$ indicates terms independent of x_{t-1} or of variables occurring after time $t - 1$. With this substitution in (8.1) we have the expression

$$\theta P_t(x_t, W_t) = \min_{x_{t-1}} [d_t + [(x - \tilde{x})'\tilde{V}^{-1}(x - \tilde{x})]_{t-1}] + \cdots \tag{8.9}$$

However, this further operation will yield exactly the recursions of the risk-neutral case; c.f. (8.2). The modified recursions asserted in the theorem are thus established. As in Section 7.3, condition (8.7) is necessary for validity of the extremum operation in (8.9). □

The quantities $\tilde{x}_{t-1}$ and $\tilde{V}_{t-1}$ are to be regarded as updatings of $\hat{x}_{t-1}$ and V_{t-1} which take account of the cost pressures at time $t - 1$. The further updating to $\hat{x}_t$ and V_t then takes account of the new observation y_t.

However, the combined effect of these two updatings is not particularly obvious if we take the risk-neutral recursions in their original forms of Section 4.2. The alternative forms of Section 4.4 yield a much more transparent result.

8.2 ALTERNATIVE FORMS OF THE RECURSIONS: COST PRESSURE AS INFORMATION

Theorem 8.2.1. *Assume L normalized to zero. Then the recursion for V_t can be written in the form*

$$V_t = N + A(V_{t-1}^{-1} + I(\theta))^{-1} A' \qquad (8.10)$$

where

$$I(\theta) = C'M^{-1}C + \theta R \qquad (8.11)$$

The recursion for $\hat{x}_t$ (the Kalman filter) can be written in the form

$$\hat{x}_t = A\hat{x}_{t-1} + Bu_{t-1} + \alpha_t + H_t(y_t - C\hat{x}_{t-1}) - \theta A(V_{t-1}^{-1} + I(\theta))^{-1}(R\hat{x}_{t-1} + S'u_{t-1}) \qquad (8.12)$$

where

$$H_t = A(V_{t-1}^{-1} + I(\theta))^{-1} C' M^{-1} \qquad (8.13)$$

If L has not been normalized to zero then A, N and α_t must be substituted by $A - LM^{-1}C$, $N - LM^{-1}L'$ and $\alpha_t + LM^{-1}y_t$, respectively.

Proof. The evaluations (8.10) and (8.13) for V_t and H_t follow immediately from expressions (4.16) and (4.17) if we make the substitution of $\tilde{V}_{t-1}^{-1} = V_{t-1}^{-1} + \theta R$ for V_{t-1}^{-1}. By Theorem 8.1.1 the Kalman filter will take the form

$$\hat{x}_t = A\tilde{x}_{t-1} + Bu_{t-1} + \alpha_t + H_t(y_t - C\tilde{x}_{t-1})$$
$$= A\hat{x}_{t-1} + Bu_{t-1} + \alpha_t + H_t(y_t - C\hat{x}_{t-1}) + (A - H_tC)(\tilde{x}_{t-1} - \hat{x}_{t-1}) \qquad (8.14)$$

We readily find from expressions (8.13) for H_t and (8.6) for $\tilde{x}_{t-1}$ that the final term in (8.14) reduces to the form indicated in (8.12). See Exercise 4.1.1 for achievement of the normalization of L to zero. □

We see that the principal effect (8.11) of risk sensitivity is to modify the information matrix of an observation, $C'M^{-1}C$ to the value $C'M^{-1}C + \theta R$. That is, there is a 'cost-pressure' on estimates which manifests itself in an apparent increase in the quality of observation if θ is positive, an apparent decrease if θ is positive. This is, of course, completely parallel to the apparent augmentation of control power by noise power noted in Section 7.4.

The other effect of risk sensitivity is the appearance of the final term in the Kalman filter (8.12). This tends to bias $\hat{x}_t$ in a cost-decreasing or -increasing

110   THE MARKOV CASE: PAST STRESS AND ESTIMATION

direction, according to the sign of θ, and is another manifestation of optimism or pessimism.

The analogue of Γ_t, the actual gain matrix of the control case, is

$$\Omega_t = A - H_t C = A(V_{t-1}^{-1} + I(\theta))^{-1}(V_{t-1}^{-1} + \theta R) \qquad (8.15)$$

The analogue of $\tilde{\Gamma}_t$, the predictive gain matrix, is the whole coefficient of $\hat{x}_{t-1}$ in (8.12), which is

$$\tilde{\Omega}_t = A - H_t C - \theta A(V_{t-1}^{-1} + I(\theta))^{-1} = A(V_{t-1}^{-1} + I(\theta))^{-1} V_{t-1}^{-1} = \Omega_t (I + \theta V_{t-1} R)^{-1} \qquad (8.16)$$

and it is $\tilde{\Omega}_t$ which we would expect to converge to a stability matrix with increasing t.

8.3 RECOUPLING

It is the evaluation of $\check{x}_t$ as the value of x_t θ-extremizing $P_t(x_t, W_t) + F_t(x_t)$ which recouples the calculations, and provides the actual MS evaluations of estimates and controls at time t. We now have solutions.

$$F_t(x_t) = [x' \Pi x - 2\sigma' x]_t + \cdots \qquad (8.17)$$

$$P_t(x_t, W_t) = \theta^{-1}[(x - \hat{x}) V^{-1} (x - \hat{x})]_t + \cdots \qquad (8.18)$$

for the two stresses.

Theorem 8.3.1. *Suppose $V_t^{-1} + \theta \Pi_t$ positive definite. Then the minimum stress estimate of x_t at time t is*

$$\check{x}_t = (I + \theta V_t \Pi_t)^{-1} (\hat{x}_t + \theta V_t \sigma_t) \qquad (8.19)$$

Proof. The θ-extremizing value of x_t will be the value minimizing $\theta(P_t + F_t)$, which is a quadratic form in x_t with matrix $V_t^{-1} + \theta \Pi_t$. For propriety we require that this matrix be positive-definite. The minimizing value is easily determined as expression (8.19). □

The recoupling expressed in equation (8.19) in fact marks a crowning moment, although it may not seem spectacular. We have seen the separation principle hold, in that the P- and F-recursions could be solved separately by methods which are familiar, apart from the addition of the spice of risk effects. Recoupling now finds its expression in the evaluation of estimate (8.19), and certainty equivalence in the fact that the optimal control is given by $u_t = u_t(\check{x}_t)$, where $u_t(x_t)$ is the perfect-observation control, given by (7.47).

We again note some asymmetry in the P and F evaluations, however, in that expressions (8.17) and (8.18) have matrices Π_t and V_t^{-1}, respectively. We could achieve full symmetry by adapting the approach of Chapter 5 to the risk-sensitive

8.5 CONTINUOUS TIME

case. However, the path-integral techniques of Part III incorporate this reformulation naturally; see, in particular, Section 10.3.

8.3.1 Exercises and comments

(1) The conventions on the way the linear term is introduced may seem different in (8.17) and (8.18). However, this is again a reflection of the incomplete symmetry of the formulation. We know from Chapter 5 that that the quantities $\hat{x}_t$ and σ_t are in fact mutual analogues in the symmetric formulation.

(2) Assume that F is rewritten as in Exercise 7.6.1. Relation (8.19) then becomes

$$\check{x}_t = (I + \theta V_t \Pi_t)^{-1}(\hat{x}_t + \theta V_t \Pi_t x_t^*)$$

which makes the MSE $\check{x}_t$ an evident average of the values $\hat{x}_t$ and x_t^* extremizing past and future stress.

8.4 PROPER RANGES OF θ

All calculations performed in the last two sections are proper as long as the current V is positive definite. If calculations fail at any time (because the form being minimized in (8.1) is not positive definite) then they fail at all subsequent times. Thus, if $\bar{\theta}_t$ is the infimal value of θ for which calculations remain proper up to time t, then $\bar{\theta}_t$ is non-decreasing in t.

A particular case which is easily analysed is that for which $N = 0$, $A = I$, when $x_t = x_0$ for all t. Relation (8.10) for V_t then has the immediate solution

$$V_t = (V_0^{-1} + tI(\theta))^{-1}$$

with $I(\theta) = C'M^{-1}C + \theta R$. The value of $\bar{\theta}_t$ is then the greatest value of θ which makes $V_0^{-1} + tI(\theta)$ singular. This indeed increases with t, to the greatest value $\bar{\theta}$ which makes $I(\theta)$ singular.

8.5 CONTINUOUS TIME

In continuous time either form of the Riccati equation becomes

$$\dot{V} = N + AV + VA' - (L + VC')M^{-1}(L' + CV) - \theta VRV \qquad (8.20)$$

This reduces to

$$\dot{V} = N + AV + VA' - V(C'M^{-1}C + \theta R)V \qquad (8.21)$$

if L has been normalized to zero. (The transformations of A, N and α under this normalization listed at the end of Theorem 8.2.1 remain valid for the continuous-time case.)

The Kalman filter (8.12) becomes

$$\dot{\hat{x}} = A\hat{x} + Bu + \alpha + H(y - C\hat{x}) - \theta V(R\hat{x} + S'u) \qquad (8.22)$$

with

$$H = (L + VC')M^{-1} \tag{8.23}$$

Recoupling is achieved exactly as for the discrete-time case: the MSE $\check{x}_t$ is given by (8.19) and the optimal control by $u_t(\check{x}_t)$.

8.6 NOTES ON THE LITERATURE

Jacobson, in his pioneering work (1973, 1977), was not able to analyse the case of imperfect state observation, in the sense of reducing the problem to finite-dimensional recursions. Such reductions were found in the particular cases of zero plant noise and of zero state costs by Speyer *et al.*, (1974), Speyer (1976) and Kumar and van Schuppen (1981). However, the general case was solved first in Whittle (1981), working from the key result of the risk-sensitive certainty-equivalence principle.

Bensoussan and van Schuppen (1984) considered the corresponding continuous-time version. This paper, while referring to Whittle (1981), gave the impression that it was announcing a first solution, an impression grudgingly corrected in Bensoussan and van Schuppen (1985). It is true that these authors use different methods. They do not, for example, establish a certainty-equivalence principle.

CHAPTER 9

The infinite horizon: limits, breakdown points and policy improvement

We would wish to establish conditions for horizon stability, i.e. conditions under which limits of relevant quantities exist as horizon or history becomes infinite. The analysis now has the extra element that results will depend upon the value of θ. There is particular interest in establishing the bounds on θ at which there is failure in optimization, to be construed as the onset of either utter despair or boundless euphoria.

We shall consider only the case of future stress; that of past stress is in complete and evident analogy. We shall assume, for convenience, that the problem has been normalized so that S is zero. The general case can be restored by substituting $A - BQ^{-1}S$ and $R - S'Q^{-1}S$ for A and R in the formulae.

9.1 THE SCALAR CASE

One gains considerable insight by examining first the scalar case, when $m = n = 1$. Let us write the compound operator $\tilde{f}f$ as χ, so that the Π-recursion (7.27) is written $\Pi_t = \chi \Pi_{t+1}$, or

$$\Pi_{(s)} = \chi \Pi_{(s-1)} \qquad (s > 0) \tag{9.1}$$

In the scalar case we have

$$\chi \Pi = R + \frac{A^2 \Pi}{1 + J\Pi} \tag{9.2}$$

where J is the effective control-power index:

$$J = J_0 + \theta N = B^2 Q^{-1} + \theta N \tag{9.3}$$

Equation (9.1) can be solved graphically by the usual construction; see Figure 5. What we shall certainly require for existence of infinite-horizon limits is that equation (9.1) should have a finite equilibrium solution Π, stable in that $d(\chi\Pi)/d\Pi < 1$ at this value, and appropriate in that all $\Pi_{(0)}$ of interest lie in the domain of attraction of this equilibrium.

Let us avoid degenerate cases by assuming R and N strictly positive. The

THE INFINITE HORIZON

problem can then equally well be parametrized in terms of varying J as of varying θ.

Theorem 9.1.1. *The infimum of J values at which equation (9.1) has a stable equilibrium solution is $J = 0$ if $|A| \geq 1$, or $J = -(1 - |A|)^2/R$ if $|A| < 1$. The corresponding infimal values of θ are then*

$$\theta_1 = -\frac{1}{N}\frac{B^2}{Q} \tag{9.4}$$

$$\theta_2 = -\frac{1}{N}\left[\frac{B^2}{Q} + \frac{(1-|A|)^2}{R}\right] \tag{9.5}$$

corresponding to the cases when the uncontrolled system is not and is strictly stable, respectively.

Proof. For $J > 0$ the graph of $\chi\Pi$ with Π rises concavely to a limit value, as shown in Figure 5. There is a clear stable positive equilibrium point, finite and unique.

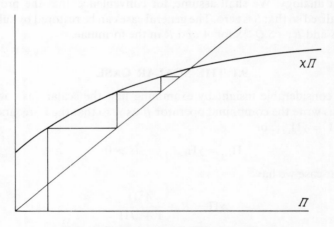

Figure 5. *Value iteration for Π in the case $J > 0$*

In the case $J = 0$ the function χ has a simple linear form, and we have the two cases of Figure 6. If $|A| < 1$ (when the uncontrolled system is strictly stable) then equation (9.1) still has a finite, stable equilibrium point. If $|A| \geq 1$ then there is no such solution, and $\Pi_{(s)} \to +\infty$ with increasing s.

In the stable case $|A| < 1$ we can decrease J somewhat further. For J negative $\chi\Pi$ is increasing and convex. If J is small enough numerically the equation $\Pi = \chi\Pi$ will have positive roots; the two roots $\Pi^{(1)}$ and $\Pi^{(2)}$ illustrated in Figure

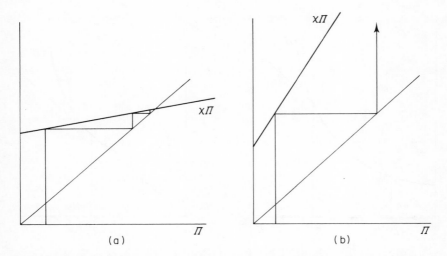

Figure 6. *Value iteration for Π when $J = 0$ in the cases (a) of a stable system ($|A| < 1$) and (b) of an unstable system ($|A| \geq 1$)*

7(a). The lower root $\Pi^{(1)}$ is stable, in that $\Pi_{(s)}$ will converge to $\Pi^{(1)}$ from any $\Pi_{(0)}$ in $[0, \Pi^{(2)})$. The root $\Pi^{(2)}$ is unstable: for $\Pi_{(0)} > \Pi^{(2)}$ we have $\Pi_{(s)} \to +\infty$.

However, a stable root does exist as J decreases, right up to the point where the two roots coincide, as illustrated in Figure 7(b). This occurs for the first value of J for which

$$(1 - RJ - A^2)^2 + 4RJ = 0$$

as J is decreased from zero. This quadratic has roots $J = -(1 \pm A)^2/R$, so the critical value of J is $-(1 - |A|)^2/R$, as asserted. □

The interpretation would be that infinite-horizon limits exist so long as $J > 0$ (i.e. $\theta > \theta_1$), because there is positive effective control power. When this no longer holds, then no faith can be placed in control, and infinite-horizon limits exist only if the system is known to be stable in the absence of control. However, even this knowledge will be insufficient to allay the fears of a real pessimist if either (1) the terminal cost is pitched high enough, or (2) θ falls below the value θ_2, implying that even the intrinsic stability of the system is no longer a convincing enough guarantee.

However, the requirement that equation (9.1) should have a finite, stable equilibrium solution Π is not the only condition. There is also the condition of risk resistance, which requires that

$$\Pi^{-1} + \theta N > 0 \tag{9.6}$$

(see equation (7.28)). Better expressed, the class of non-negative Π satisfying (9.6)

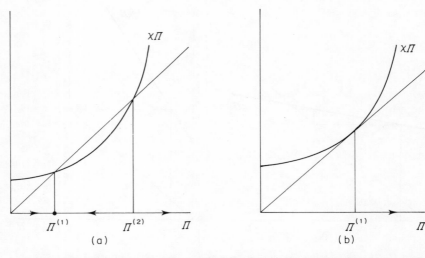

Figure 7. *Solution for Π in the case of a stable model ($|A| < 1$) with negative J. The arrow on the Π axis shows the direction of movement under value iteration. In case (a) there is still an interval of Π from which value iteration converges to the stable solution $\Pi^{(1)}$. Case (b) is the marginal case; if J is any more negative then there is no solution to the Riccati equation, and Π will become indefinitely large under value iteration*

should not be empty, and this class should contain the appropriate equilibrium solution and all values $\Pi_{(s)}$ which occur during passage to this equilibrium.

Theorem 9.1.2. *In the scalar case the risk-resistance condition requires that if*

$$A^2 - |A| \geq -B^2 R/Q \tag{9.7}$$

then θ must exceed θ_3, where

$$\theta_3 = -\frac{1}{N}[R + A^2 Q/B^2]^{-1} \tag{9.8}$$

Proof. Change variable from Π to $\beta = \Pi^{-1} + \theta N$. The equation $\Pi = \chi\Pi$ then becomes

$$\beta = \psi(\beta) + \theta N \tag{9.9}$$

where

$$\psi(\beta) = \frac{\beta + J_0}{R(\beta + J_0) + A^2}$$

(Recall that $J_0 = B^2/Q$.) We require that the appropriate solution of (9.9) should never fall as low as zero. One verifies easily that the function $\psi(\beta)$ is increasing

9.1 THE SCALAR CASE

and concave with

$$\psi'(0) = \frac{A^2}{(RJ_0 + A^2)^2} \tag{9.10}$$

If expression (9.10) does not exceed unity (which is just condition (9.7)) then equation (9.9) has a single positive solution, which decreases to zero as θ decreases to the value θ_3. If expression (9.10) is greater than unity (i.e. if (9.7) does not hold) then, as θ decreases, equation (9.9) first develops a pair of positive solutions (the greater being the relevant one) and then has no real solution at all for $\theta < \theta_2$. There can thus be no loss of risk resistance in this case. $\square$

There are no other conditions to be invoked, so we can now characterize the breakdown value of θ exactly.

Theorem 9.1.3. *If applicable, bound (9.8) supersedes those previously stated, in that $\theta_3 \geq \theta_1$ if $|A| \geq 1$ and $\theta_3 \geq \theta_2$ always.*

The exact breakdown value of θ is then θ_3 (as given by (9.8)) if inequality (9.7) holds and θ_2 (as given by (9.5)) otherwise.

Proof. We have $\theta_3 > -J_0/(A^2 N)$, which is not less than θ_1 if $|A| \geq 1$. Direct calculations show that

$$\theta_2/\theta_3 = 1 + (A^2 - |A| + RJ_0)^2/(RJ_0) \geq 1$$

whence $\theta_3 \geq \theta_2$. The first assertion of the theorem is thus established, and the second follows. $\square$

In Chapter 11 we shall see θ_3 characterized rather as the largest value of θ for which the matrix

$$\begin{bmatrix} \Pi & I \\ I & -\theta N \end{bmatrix}$$

is singular. This characterization avoids mention of inverses of Π or N, which may have to be understood in some formal sense.

9.1.1 Exercises and comments

(1) *Continuous time limits.* In continuous time the analogue of (9.1) is

$$\mathring{\Pi} = R + A'\Pi + \Pi A - \Pi J \Pi$$

In the scalar case the analogues of the critical values of Theorem 9.1.1 are

$$\theta_1 = -B^2/NQ$$

if $A \geq 0$ (i.e. if the uncontrolled system is unstable) and

$$\theta_2 = -N^{-1}[(B^2/Q) + (A^2/R)]$$

if $A < 0$ (i.e. if the uncontrolled system is stable).

(2) *Continuous time continued.* We know from Section 7.7 that the risk-resistance condition has no force in continuous time. This also follows from the limit form of Theorem 9.1.2. In the continuous time limit the applicability condition (9.7) becomes $A \geq 0$, and θ_3 becomes equal to θ_2. Thus, Theorem 9.1.2 supplies no extra condition.

(3) Note that the condition $\theta > \theta_3$ can be expressed

$$\frac{B^2}{Q} + \frac{A^2}{(\theta N)^{-1} + R} > 0,$$

a form we shall encounter in Lemma 11.5.3.

(4) Adapt the discussion of this section to the case when $R < 0$ and we expect that $\Pi < 0$. In this case the breakdown values of θ will be *upper* bounds, marking the onset of euphoria rather than of despair.

9.2 INFINITE-HORIZON LIMITS IN GENERAL

Theorem 7.4.1 contained a key observation: that the effect of risk sensitivity on the Riccati equation is simply to modify the control-power matrix $J_0 = BQ^{-1}B'$ to $J_0 + \theta N$. This observation enables us to adapt the risk-neutral treatments very easily, to obtain sufficient conditions for the existence of infinite-horizon limits in the risk-sensitive case.

Let us in fact assume θ bounded below by the 'controllability' condition in analogue to that of Theorem 3.4.1: that J be non-negative definite, and positive definite on $\{(A')^t\}$. The analysis of the scalar case, when this condition becomes simply that $J > 0$, would seem to indicate that this is a conservative condition, but could be a necessary condition if the uncontrolled system were unstable. We also know that it may not be sufficient, in that it does not ensure risk resistance.

However, the assumption does imply that the Riccati equation

$$\Pi_{(s+1)} = \tilde{ff} \Pi_{(s)} = \chi \Pi_{(s)} \tag{9.11}$$

is exactly what it would have been in a risk-neutral problem with control-power matrix J instead of J_0, and we can simply take over the analysis of Theorem 3.4.1.

Theorem 9.2.1. *Assume the Markov assumptions* (M1–4)' *of Section 4.1, re-listed in Section 7.1, and suppose S normalized to zero. Suppose also that Q is positive definite, that R is positive definite on* $\{A^t\}$, *and that* $J = BQ^{-1}B' + \theta N$ *is non-negative definite, and is positive definite on* $\{(A')^t\}$.

Then infinite-horizon limits exist, in that the Riccati equation (9.11) *has a unique non-negative definite equilibrium solution* Π, *to which* $\Pi_{(s)}$ *converges from any*

9.2 INFINITE-HORIZON LIMITS IN GENERAL

non-negative definite $\Pi_{(0)}$. The limit predictive gain matrix

$$\tilde{\Gamma} = (I + J\Pi)^{-1}A \tag{9.12}$$

is a stability matrix.

However, these conclusions are not enough. They imply that the iteration of the compound operator $\tilde{ff}$ produces a convergent result, but it is also necessary that application of the single operator $\tilde{f}$ should be meaningful at all stages; it is this which leads to the risk-resistance condition. That is, that

$$\Pi^{-1} + \theta N > 0 \tag{9.13}$$

for all relevant Π. For given θ, this sets an upper bound on admissible Π.

Let us define $\mathcal{R}$ as the set of non-negative definite matrices Π for which (9.13) holds and

$$\chi\Pi < \Pi$$

We shall certainly need to demand that $\mathcal{R}$ be non-empty, since it must contain at least the limit matrix, the equilibrium solution of (9.11).

Let us also define $\bar{\mathcal{R}}$ as the set of non-negative definite matrices bounded above by some member of $\mathcal{R}$ (in the ordering of positive definiteness).

Lemma 9.2.2. *$\mathcal{R} \subset \bar{\mathcal{R}}$, and all members of $\bar{\mathcal{R}}$ satisfy the risk-resistance condition (9.13). Furthermore, $\chi\mathcal{R} \subset \mathcal{R}$ and $\chi\bar{\mathcal{R}} \subset \bar{\mathcal{R}}$.*

Proof. The first two assertions are obvious. For the others, assume that $\Pi \in \mathcal{R}$. Since $\chi\Pi \leq \Pi$ then $\chi\Pi$ satisfies the risk-resistance condition (9.13) if Π does. Since χ is monotonic we have also that $\chi(\chi\Pi) \leq \chi\Pi$. Thus $\chi\Pi \in \mathcal{R}$, and so $\chi\mathcal{R} \subset \mathcal{R}$.

Assume that $\bar{\Pi} \in \bar{\mathcal{R}}$, so that $\bar{\Pi} \leq \Pi$ for some Π in $\mathcal{R}$. Then $\chi\bar{\Pi} \leq \chi\Pi \leq \Pi$. Thus $\chi\bar{\Pi}$ is bounded above by an element of $\mathcal{R}$, and so belongs to $\bar{\mathcal{R}}$. □

Therefore, if we choose $\Pi_{(0)}$ in $\bar{\mathcal{R}}$, then $\Pi_{(s)} = \chi^s \Pi_{(0)}$ also belongs to $\bar{\mathcal{R}}$, and so is 'legitimate' for all s. We thus conclude:

Theorem 9.2.3. *Add to the hypotheses of Theorem 9.2.2 that $\mathcal{R}$ is non-empty. Then, for $\Pi_{(0)}$ in $\bar{\mathcal{R}}$ the sequence $\Pi_{(s)} = \chi^s \Pi_0$ obeys the risk-resistance condition for all s, as does then the limit matrix Π.*

The two conditions, on J and $\mathcal{R}$, place a sufficient lower bound on θ. We do not have an explicit evaluation of this lower bound, and the bound may also be conservative, in that it is far from sharp. However, the positive-definiteness condition on J is a simple one, in that it leads to a simple transference of the risk-neutral results.

9.3 ANALYSIS OF A VECTOR EXAMPLE

We shall consider a vector example which is explicitly soluble, and which is found to have an independent interest later (see Section 13.3).

Consider the scalar problem with plant equation

$$x_t = ax_{t-1} + b_1 u_{t-1} + b_2 u_{t-2} + \varepsilon_t \tag{9.14}$$

so that the control has an effect at both one and two lags. We take the problem as risk sensitive, and so could bring it to a Markov risk-neutral form by defining state and control vectors

$$\vec{x}_t = \begin{bmatrix} x_t \\ u_{t-1} \end{bmatrix} \quad \vec{u}_t = \begin{bmatrix} u_t \\ \varepsilon_{t+1} \end{bmatrix}.$$

Let us distinguish the matrices of this standardized formulation also by the vector superscript. We have then

$$\vec{A} = \begin{bmatrix} a & b_2 \\ 0 & 0 \end{bmatrix} \quad \vec{B} = \begin{bmatrix} b_1 & 1 \\ 1 & 0 \end{bmatrix}$$

$$\vec{R} = \begin{bmatrix} R & 0 \\ 0 & 0 \end{bmatrix} \quad \vec{Q} = \begin{bmatrix} Q & 0 \\ 0 & (\theta N)^{-1} \end{bmatrix}$$

and we assume S (and so $\vec{S}$) zero. We have then

$$\vec{J} = \begin{bmatrix} b_1^2 Q^{-1} + \theta N & b_1 Q^{-1} \\ b_1 Q^{-1} & Q^{-1} \end{bmatrix} \tag{9.15}$$

$$\vec{J} + \vec{A}\vec{J}\vec{A}' = \vec{J} + \begin{bmatrix} (ab_1 + b_2)^2 Q^{-1} + \theta N a^2 & 0 \\ 0 & 0 \end{bmatrix} \tag{9.16}$$

One can write down the Riccati equation for the 2×2 matrix $\vec{\Pi}$ directly. When one does so, one finds that the matrix has the form

$$\vec{\Pi} = \begin{bmatrix} R + a^2 \beta & ab_2 \beta \\ ab_2 \beta & b_2^2 \beta \end{bmatrix}$$

where β is a scalar constant. This reflects the fact that the value function is of the form

$$(\vec{x}' \vec{\Pi} \vec{x})_t = R x_t^2 + \beta(ax_t + b_2 u_{t-1})^2 \tag{9.17}$$

which, in turn, follows from the fact that the scalar variable

$$w_t = ax_t + b_2 u_{t-1} \tag{9.18}$$

follows an autonomous first-order equation

$$w_t = aw_{t-1} + (ab_1 + b_2)u_{t-1} + a\varepsilon_t \tag{9.19}$$

If one writes down the optimality equation afresh one finds that the value function indeed has the form (9.17), with β determined by

$$\beta w^2 = \min_u \text{ext}_\varepsilon \; [R(w + b_1 u + \varepsilon)^2 + Qu^2 + (\theta N)^{-1}\varepsilon^2$$
$$+ \beta(aw + (ab_1 + b_2)u + a\varepsilon)^2] \quad (9.20)$$

The constant β thus obeys the equation

$$(ab_1 + b_2)^2 \beta^2 + [Q(1 - a^2) + R(b_1^2 - b_2^2)]\beta - QR$$
$$+ N\theta[(Qa^2 + Rb_2^2)\beta^2 + QR\beta] = 0 \quad (9.21)$$

Consider first the risk-neutral case $\theta = 0$. As in Section 9.1, the problem becomes uncontrollable when the coefficient of β^2 in (9.21) is zero, i.e. $ab_1 + b_2 = 0$. We see from (9.19) that it is exactly the component w_t which then becomes uncontrollable. There should still be an acceptable solution for β if the uncontrolled process is stable, and we find from (9.21) that the solution is

$$\beta = \frac{QR}{(Q + Rb_1^2)(1 - a^2)} \quad (9.22)$$

acceptable if $|a| < 1$.

In the risk-sensitive case we require for effective controllability that the coefficient of β^2 in (9.21) be positive, i.e. that

$$\theta > -\frac{1}{N} \frac{(ab_1 + b_2)^2}{Qa^2 + Rb_2^2} \quad (9.23)$$

This bound is comparable with the bound θ_1 of (9.4). The naive demand for effective controllability, which should give a conservatively high bound on θ, is that $\vec{J} \geq 0$ and $\vec{J} + \vec{A}\vec{J}\vec{A}' > 0$ (see Exercise 9.3.1). This yields the indeed conservative evaluation $\theta \geq 0$.

As in Section 9.1, bound (9.23) may be superseded under some conditions by the risk-resistance condition. For θ negative this latter will require that the coefficient of ε^2 in the right-hand member of expression (9.20) be negative, i.e. that

$$(\theta N)^{-1} + R + a^2 \beta < 0 \quad (9.24)$$

9.3.1 Exercises and comments

(1) The criterion for effective controllability must include $\vec{J} \geq 0$. If we simply require that $\vec{J} + \vec{A}\vec{J}\vec{A}' > 0$ we obtain

$$\theta > -\frac{1}{N} \frac{(ab_1 + b_2)^2}{Q(1 + a^2)} \quad (9.25)$$

There are circumstances ($Rb_2^2 > Q$) under which the true bound (9.23) falls below the supposedly conservative bound (9.25).

(2) An even simpler case is that in which we assume $a = b_1 = 0$, so that (9.23) becomes $\theta > -(NR)^{-1}$. As a reduced version of (9.17) we have a value function of the form $Rx_t^2 + \gamma u_{t-1}^2$. The optimality equation reduces to

$$\gamma u_{t-1}^2 = \min_{u_t} \text{ext}_{\varepsilon_{t+1}} [R(b_2 u_{t-1} + \varepsilon_{t+1})^2 + Q u_t^2 + (\theta N)^{-1} \varepsilon_{t+1}^2 + \gamma (b_2 u_t)^2]$$

We then have $u_t = 0$ and the condition (for θ negative)

$$(\theta N)^{-1} + R < 0$$

provides both the risk-resistance condition and the effective controllability condition. Of course, it also imples the known lower bound $-(NR)^{-1}$, and we have $\gamma = b_2^2 R(1 + \theta NR)^{-1}$. This corresponds to $\beta = R(1 + \theta NR)^{-1}$; the other solution of (9.21) is $\beta = -Qb_2^{-2}$.

9.4 POLICY IMPROVEMENT

Assume that $x'\Pi x$ is the infinite-horizon cost function in the risk-neutral case for some control policy. It follows then from definitions that $f\Pi \leq \Pi$, which gives one a mechanism for policy improvement. Moreover, assume that $f\Pi = f_\Pi \Pi$, where f_Π is the linear operator associated with the revised policy determined by the evaluation of $f\Pi$. That is, for two matrices, Π, Π^*

$$f_\Pi \Pi^* = R + K'S + S'K + K'QK + (A + BK)'\Pi^*(A + BK)$$

where

$$K = -(Q + B'\Pi B)^{-1}(S + B'\Pi A)$$

Then the standard theory, which proves that policy improvement produces a sequence of policies for which the matrix Π will decrease to the optimal value, relies on two properties of f_Π. These are, that f_Π in monotonic, and that $f\Pi^* \leq f_\Pi \Pi^*$.

In the risk-sensitive case we have the double control (u, ε) to be determined. However, the specification of a policy can only amount to the specification of a rule determining u. When we determine $E_\pi \exp(-\tfrac{1}{2}\theta \mathbb{C})$ then the effective value of ε is determined by the integration with respect to noise ε which is implied in the expectation. This then means that ε is determined by an extremal operation. In other words, whatever policy one chooses, optimal or not, Nature always optimizes on her own account.

Nevertheless, if we are to carry over the idea of policy improvement to the risk-sensitive case in a simple form we would wish to behave as though the rules for the determination of both u and ε could be laid down, and then improved. In order to achieve this we shall have to demonstrate that the operator $\chi = f\tilde{f}$ has the essential properties shown by f in the risk-neutral case. These are that

$$\chi \Pi \leq \Pi \tag{9.26}$$

9.4 POLICY IMPROVEMENT

for a suitably large class of matrices Π, and that χ_Π is monotonic (this being, as for f_Π, the linear operator determined by $\chi\Pi = \chi_\Pi\Pi$).

We solve the first point by fiat: we postulate that $\mathcal{R}$ is non-empty, where (be it recalled from Section 9.2) $\mathcal{R}$ is the class of non-negative-definite matrices Π satisfying (9.26) and the risk-resistance condition

$$\Pi^{-1} + \theta N > 0 \tag{9.27}$$

It is certainly necessary that $\mathcal{R}$ be non-empty, for at least the optimal Π (if one exists) must belong to it. In Theorem 9.4.4 we shall demonstrate a method for generating members of $\mathcal{R}$.

We assume throughout the chapter that the positive-definiteness conditions on the cost function of condition (M3), Section 3.2, hold. There is then no problem with the case $\theta \geq 0$. Condition (9.27) will hold trivially. The operator χ_Π will be monotonic, and, if Π is the matrix for the cost function of some policy which specifies both u and ε, then inequality (9.26) will hold. This is because the situation in the combined pair of control variables (u, ε) is exactly as it is in the risk-neutral case of a single control u. It is the case $\theta < 0$, when ε works against u, which may prove troublesome.

Lemma 9.4.1. *Assume that $J \geq 0$ and that Π, Π^* are two non-negative definite $n \times n$ matrices. Then*

$$\chi\Pi^* \leq \chi_\Pi\Pi^* \tag{9.28}$$

Proof. One can give a direct manipulative proof (see Exercise 9.4.1), but there is an 'as though' proof which is much more economical. Relation (9.28) holds by definition of the operator χ in the case $\theta \geq 0$, because then χ minimizes with respect to both u and ε. This is no longer obviously so in the case $\theta < 0$. However, as long as $J \geq 0$, the calculations can be regarded as though they pertained to a conventional risk-neutral case, with control-power matrix J. Thus (9.28) holds. $\square$

Theorem 9.4.2. *In the case $\theta \geq 0$ policy improvement works as in the risk-neutral case.*

In the case $\theta < 0$, assume $J \geq 0$ and $\mathcal{R}$ non-empty. Then policy improvement works in that, if applied to a member of $\mathcal{R}$, it produces a non-increasing sequence of matrices in $\mathcal{R}$, whose limit lies in $\mathcal{R}$ and is optimal (i.e. is the matrix of the value function).

Proof. The first statement follows from the discussion before the theorem. By the definition of $\mathcal{R}$ we know that $\chi\Pi \leq \Pi$, and we know from Theorem 9.2.1 that equality implies optimality. Consider then the case $\Pi > \chi\Pi = \chi_\Pi\Pi$. The linear operator χ_Π is monotonic, so $\Pi^*_{(s)} = \chi_\Pi^s\Pi$ is non-increasing in s, and so converges to a limit $\Pi^* < \Pi$, by the usual arguments.

Now, since Π^* is less than Π, it will satisfy the risk-resistance condition, since Π does. Furthermore, appealing to Lemma 9.4.1 we have

$$\chi\Pi^* \leq \chi_\Pi \Pi^* = \Pi^*$$

so that Π^* obeys the deficiency condition (9.26). Thus $\Pi^* \in \mathscr{R}$. One stage of policy improvement has thus yielded a smaller matrix, also in $\mathscr{R}$. The conclusions of the theorem then follow. □

Therefore policy improvement works even for negative θ, provided $J \geq 0$ and we start with a matrix in the class $\mathscr{R}$.

We know how to generate elements of $\mathscr{R}$ in the risk-neutral case: as the matrices Π of the cost function $x'\Pi x$ for any stabilizing, stationary, linear Markov policy. The following theorem gives us an analogous construction for the risk-sensitive case.

Theorem 9.4.3. *Assume that $J \geq 0$. Assume also that the policy (i.e. the rule for both u and ε) determined by the application of χ to a matrix Π is stabilizing, in that the linear equation for Π^**

$$\chi_\Pi \Pi^* = \Pi^*$$

has a finite solution. Then

$$\chi\Pi^* \leq \Pi^*$$

Proof. We appeal simply to Lemma 9.4.1, and have

$$\chi\Pi^* \leq \chi_\Pi \Pi^* = \Pi^*$$ □

We see then that Π^* has the first qualification, (9.26), to belong to $\mathscr{R}$. It may not have the second: the risk-resistance property (9.27). However, the policy-improvement routine starting from Π^* will generate a non-increasing sequence of matrices satisfying (9.26), and which ultimately satisfy (9.27) if $\mathscr{R}$ is non-empty.

As in Section 3.8, the method of policy improvement is equivalent to the application of the Newton–Raphson algorithm to the equation $\Pi = \chi\Pi$. One can see from graphical examination of the scalar case that limitation of the initial Π to values in $\mathscr{R}$ may be conservative. In this case, the set $\mathscr{R}$ is just the interval $[\Pi_{eq}, -(\theta N)^{-1}]$, where Π_{eq} is the equilibrium value; see Figure 8. The lower bound is set by (9.26) and the upper by (9.27). In fact, the Newton–Raphson algorithm would converge from any Π exceeding Π_{low}, the point at which $\chi\Pi$ has unit derivative.

9.4.1 Exercises and comments

(1) The control rule induced by calculation of $\chi\Pi$ is

$$u = -Q^{-1}B'(\Pi^{-1} + J)^{-1}Ax$$

9.4 POLICY IMPROVEMENT

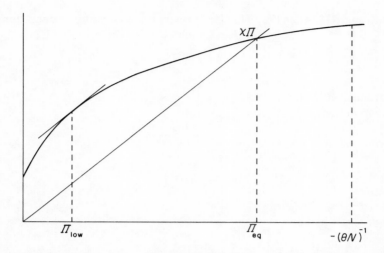

Figure 8. *The theory of this section guarantees that the policy-improvement (Newton–Raphson) algorithm will produce a sequence of Π-values converging monotonely to Π_{eq} from any initial value in $(\Pi_{eq}, -(\theta N)^{-1})$. In fact, convergence will take place from any initial value exceeding Π_{low}*

What is the determination of ε? Show that

$$\chi_\Pi \Pi^* = R + (\Pi^{-1} + J)^{-1} J (\Pi^{-1} + J)^{-1}$$
$$+ [I - (\Pi^{-1} + J)^{-1} J] \Pi^* [I - J(\Pi^{-1} + J)]$$

(2) Show that

$$\chi_\Pi \Pi^* - \chi \Pi^*$$
$$= A'(\Pi^{-1} + J)^{-1}(\Pi^{-1}\Pi^* - I)(\Pi^* + J^{-1})^{-1}(\Pi^*\Pi^{-1} - I)(\Pi^{-1} + J)^{-1} A$$

and this expression is non-negative if $\Pi^* + J^{-1} > 0$. This matrix can be given a meaning even if J is singular.

PART III

The path-integral (Hamiltonian) approach

By the 'path-integral approach' we mean the formulation of the optimal control problem, stochastic though it is, as the free extremization of a quadratic path integral—free in that there are no constraints of functional relation or functional dependence on the variables. The risk-sensitive certainty-equivalence principle allows us to do exactly that. Stationarity conditions on the path integral lead to a set of linear equations, of a symmetric or Hamiltonian form, whence the alternative description. A canonical operator factorization reduces these to a form which exhibits the optimal stationary control explicitly.

Such ideas have a long history in the control literature. Most of Part III is nevertheless new. It has hitherto not been realized that even the risk-sensitive case fits naturally into this formalism. Indeed, it will be apparent already from Section 10.3 that the extension to the LEQG class of models provides the natural completion of the formalism. It reveals structure which is only incompletely apparent for the risk-neutral LQG model, now seen as virtually a degenerate case.

CHAPTER 10

Path-integral methods: the formalism

10.1 HYPOTHESES

We now allow the higher-order dynamics already introduced in Section 1.3. In the operator notation of that section we assume plant and observation relations

$$\mathscr{A}x + \mathscr{B}u = \alpha + \varepsilon \tag{10.1}$$

$$y + \mathscr{C}x = \eta \tag{10.2}$$

Here $\alpha = \alpha_t$ is a known deterministic driving term. We shall, for the moment, continue with the assumption that $\{\varepsilon_t, \eta_t\}$ is a white-noise driving process with zero mean and covariance matrix

$$\operatorname{cov}\begin{bmatrix} \varepsilon \\ \eta \end{bmatrix} = \begin{bmatrix} N & L \\ L' & M \end{bmatrix} \tag{10.3}$$

The time argument t is understood for all variables in (10.1) and (10.2). The symbols $\mathscr{A}, \mathscr{B}, \mathscr{C}$ denote time-invariant realizable linear operators. If we are in the discrete-time case and assume pth-order dynamics then these will have the expressions

$$\mathscr{A} = \sum_{j=0}^{p} A_j \mathscr{T}^j$$

$$\mathscr{B} = \sum_{j=1}^{p} B_j \mathscr{T}^j \tag{10.4}$$

$$\mathscr{C} = \sum_{j=0}^{p} C_j \mathscr{T}^j$$

We shall assume A_0 non-singular, so that equation (10.1) represents a genuine forward recursion for x. Indeed, we shall usually assume the normalization $A_0 = I$. One assumes that B_0 is zero in order to keep the causal flow clear (i.e. so that controls cannot affect the very variables upon which they are supposed to be based). Conventionally, one assumes that $C_0 = 0$, although this is not necessary.

In the Markov case ($p = 1$) one had to specify stochastic initial conditions by assuming the distribution of x_0 conditional on W_0 to be known; in fact, normal with mean $\hat{x}_0$ and covariance matrix V_0. In the case of general dynamics one will similarly have to assume that knowledge of W_0 includes knowledge of any past actions $u_{-1}, u_{-2}, \ldots$ which appear in the plant equation for $t > 0$, and that one knows the initial distribution of any past process variables $x_0, x_{-1}, \ldots$ which appear in the plant or observation relations for $t > 0$. Let us denote these 'sufficient histories' by U_{-1} and X_0, respectively. In keeping with the general Gaussian assumption, we shall assume the distribution of the initial process history to be Gaussian, and that

$$f(X_0|W_0) \propto \exp[-\tfrac{1}{2}\mathbb{D}_0] \tag{10.5}$$

where $\mathbb{D}_0$ is a known quadratic function of X_0 and U_{-1}. These hypotheses could be relaxed; see Exercise 10.1.1.

For the moment we shall retain the same cost-functions as previously, although in uncentred form:

$$\mathbb{C} = \sum_{t=0}^{h-1} c_t + \mathbb{C}_h \tag{10.6}$$

where

$$c_t = \left[\begin{bmatrix} x - \bar{x} \\ u - \bar{u} \end{bmatrix}' \begin{bmatrix} R & S' \\ S & Q \end{bmatrix} \begin{bmatrix} x - \bar{x} \\ u - \bar{u} \end{bmatrix} \right]_t \tag{10.7}$$

$$\mathbb{C}_h = [(x - \bar{x})' \Pi (x - \bar{x})]_h \tag{10.8}$$

However, our formalism will itself suggest natural generalizations of both cost structure and noise statistics.

Finally, we adopt the risk-sensitive or LEQG criterion

$$\gamma_\pi(\theta) = -(2/\theta)\log[E_\pi \exp(-\theta\mathbb{C}/2)] \tag{10.9}$$

10.1.1 Exercise and comments

(1) Rather than assuming U_{-1} known, one need merely suppose that one has a normal joint initial distribution

$$f(X_0, U_{-1}|W_0) \propto \exp(-\tfrac{1}{2}\mathbb{D}_0)$$

where $\mathbb{D}_0$ is known quadratic function of X_0 and U_{-1}. The only difference this would make in the sequel is that U_{-1} would have to be included among the unobservables, to be estimated.

10.2 THE CASE OF PERFECT OBSERVATION: MAXIMUM AND STOCHASTIC MAXIMUM PRINCIPLES

In setting up the control-optimization problem as a path-extremization one we go through a stage when structure is somewhat obscured by details. It is, however, by going through this stage that we motivate the general approach.

10.2 THE CASE OF PERFECT OBSERVATION

With the motivation completed, one is in a position to appreciate the relative simplicity of the general structure.

Consider first the noiseless case, when the plant equation (10.1) becomes

$$\mathscr{A}x + \mathscr{B}u = \alpha \tag{10.10}$$

We also assume perfect observation in that the lagged process variable x_{t-1} is supposed observable at time t. As always, the formalism seems to work out more naturally if one assumes a lag in observation. For the case when the current process variable x_t is also assumed observable, see Theorem 10.2.2 at the end of the section.

In this totally deterministic situation risk-sensitivity is irrelevant. The criterion $\gamma_\pi(\theta)$ is simply an increasing function of cost $\mathbb{C}$, whatever θ, and the task is one of minimizing $\mathbb{C}$, given by (10.6), subject to the plant equation (10.10). If, as in Sections 3.6 and 3.7, we associate a Lagrangian multiplier vector λ_t with the constraint (10.1) at time t, then we look for stationary values of the Lagrangian form

$$\mathbb{I}(x, u, \lambda) = \sum_{\tau=0}^{h-1} c_\tau + \mathbb{C}_h + 2 \sum_{\tau=1}^{h} \lambda'_\tau (\mathscr{A}x + \mathscr{B}u - \alpha)_\tau \tag{10.11}$$

This is the path integral, to be minimized with respect to the primal x- and u-variables and maximized with respect to the dual λ-variables.

Assume we have progressed to time t, so that u_t is to be determined for $t \leq \tau < h$. Writing down the stationarity conditions for $\mathbb{I}$ with respect to $(x, u, \lambda)_\tau$ we deduce a set of equations which can be written (c.f. equation (3.77))

$$\Phi(\mathscr{T}) \begin{bmatrix} x \\ u \\ \lambda \end{bmatrix}_\tau^{(t)} = \begin{bmatrix} R\bar{x} + S'\bar{u} \\ S\bar{x} + Q\bar{u} \\ \alpha \end{bmatrix}_\tau \quad (t \leq \tau < h) \tag{10.12}$$

Here $\mathscr{T}$ is again the backwards translation operator acting on the subscript τ; the superscript (t) on the vector is just to show that these are the values appropriate for an optimization starting from time t. The symbol $\Phi(\mathscr{T})$ denotes the symmetric matrix of operators

$$\Phi(\mathscr{T}) = \begin{bmatrix} R & S' & \bar{\mathscr{A}} \\ S & Q & \bar{\mathscr{B}} \\ \mathscr{A} & \mathscr{B} & 0 \end{bmatrix} \tag{10.13}$$

where $\mathscr{A}$ and $\mathscr{B}$ are the operators defined in (10.4), and the relation of the *conjugate operator* $\bar{\mathscr{A}}$ to $\mathscr{A}$ is

$$\bar{\mathscr{A}} = \sum_{j=0}^{p} A'_j \mathscr{T}^{-j} \tag{10.14}$$

The equation system (10.12) is subject to boundary conditions at τ equal to t or

132   PATH-INTEGRAL METHODS: THE FORMALISM

h which we shall clarify when we formulate the general case in the next section; see also Theorem 10.2.2.

The point of this approach is that, as indicated in Section 3.7, the infinite-horizon optimal control is determined explicitly by a canonical factorization of the matrix $\Phi(\mathcal{T})$. We return to this central contention in Section 10.4.

This analysis is nothing but a generalization of the maximum principle (see Section 3.6) to the case of higher-order dynamics. The final relation in the partitioned equations of (10.12) is just the plant equation, describing the forward evolution of x. The second equation is the optimality equation, determining u in terms of immediately past values of x and u and immediately future values of λ. The first equation determines the backward evolution of λ.

Now assume that the plant equation has the noisy form

$$\mathcal{A}x + \mathcal{B}u = \alpha + \varepsilon \tag{10.15}$$

but, again, that the process variable is observable at unit lag. Prediction of x for the controlled process will not be perfect in this case. Let us, as in Part II, use $w^{(t)}$ to denote the minimal stress evaluation of a variable w based on information available at time t. Thus, for example, $u_\tau^{(t)}$ is the known value of u_τ if $\tau < t$, the optimal value of u_τ if $\tau = t$, and the best estimate at time t of the optimal value of u_τ if $\tau > t$.

Theorem 10.2.1. *In the case of a noisy plant equation with process variable observable at unit lag equations (10.12) still hold, but with $\Phi(\mathcal{T})$ having the revised definition*

$$\Phi(\mathcal{T}) = \begin{bmatrix} R & S' & \bar{\mathcal{A}} \\ S & Q & \bar{\mathcal{B}} \\ \mathcal{A} & \mathcal{B} & -\theta N \end{bmatrix} \tag{10.16}$$

The conjugate variable λ is related to the MSE of plant noise by

$$\theta N \lambda_\tau^{(t)} = \varepsilon_\tau^{(t)} \qquad (t \leq \tau < h) \tag{10.17}$$

Proof. By the RSCEP the optimal value of u_t is $u_t^{(t)}$, and the MSE's θ-extremize the stress

$$\mathbb{S} = \sum_{\tau=0}^{h-1} c_\tau + \mathbb{C}_h + \theta^{-1} \sum_{\tau=1}^{h} (\varepsilon' N^{-1} \varepsilon)_\tau \tag{10.18}$$

with ε determined in terms of x and u by the plant equation (10.15). If we again account for this constraint by multiplier vectors λ_t then we have the Lagrangian form

$$\sum_{\tau=0}^{h-1} c_\tau + \mathbb{C}_h + \sum_{\tau=1}^{h} [\theta^{-1}(\varepsilon' N^{-1} \varepsilon) + 2\lambda'(\mathcal{A}x + \mathcal{B}u - \alpha - \varepsilon)]_\tau$$

10.2 THE CASE OF PERFECT OBSERVATION

Extremizing with respect to ε_τ we obtain relation (10.17) and are left with the path integral

$$\mathbb{I}(x,u,\lambda) = \sum_{\tau=0}^{h-1} c_\tau + \mathbb{C}_h + \sum_{\tau=1}^{h} [2\lambda'(\mathscr{A}x + \mathscr{B}u - \alpha) - \theta\lambda'N\lambda]_\tau \quad (10.19)$$

The stationarity conditions for this form with respect to $(x, u, \lambda)_\tau$ amount just to equations (10.12) with $\Phi(\mathscr{T})$ having the revised definition (10.16). The stationarity condition with respect to λ_τ implies also the identification (10.17). □

The point of the theorem is that we see how neatly risk sensitivity has accommodated itself. The effect is simply that the empty bottom-right position in matrix (10.13) acquires an entry: $-\theta N$.

We can also say that the RSCEP has supplied us with something that has long been sought; a stochastic version of the Pontryagin maximum principle. We return to this point, which needs some appreciation, in Section 10.4. Further, the conjugate variable λ has the interpretation (10.17), relating it to the MSE of plant noise.

If the current process variable x_t has been observed at time t, then the equation system (10.12) becomes modified in a somewhat niggling fashion, which is why the assumption of lagged observation works out more naturally. One will no longer have stationarity conditions for x_t or λ_t, so equations (10.12) will hold for $t < \tau < h$, with the second relation (the stationarity condition with respect to u_t) holding also at $\tau = t$. However, there is a straightforward conclusion, which fits in with the canonical factorizations to be established later.

Theorem 10.2.2. *Consider the expression for $u_t^{(t)}$ in terms of $x_t^{(t)}$ and x_τ, u_τ ($\tau < t$) obtained by solving system (10.12) less the first and third relations at $\tau = t$, but plus terminal conditions. If the current value of process variable x_t has in fact been observed, then the optimal value of u_t is obtained by substituting x_t for $x_t^{(t)}$ in this expression.*

Proof. Note first that there is such an expression for $u_t^{(t)}$, in the case of lagged observation. Equation (10.12) contains no lagged values of λ, so we can use the terminal conditions and equations (10.12) for $\tau > t$ to solve for $\lambda_\tau^{(t)}$ ($\tau > t$) in terms of $x_\tau^{(t)}$ and $u_\tau^{(t)}$ ($\tau \leq t$). Substituting these evaluations of future λ into the second relation of (10.12) at $\tau = t$ we obtain an equation for $u_t^{(t)}$ in terms of the variables stated.

However, we have used just the equations which would have been available if x_t had been observed, in which case x_t itself would have taken the place of $x_t^{(t)}$. □

In the case of lagged observation we should have had to use the extra relations then available in order to evaluate the estimate $x_t^{(t)}$ in terms of observables.

Eliminating $\lambda_\tau^{(t)}$ from the first and third relations of (10.12) at $\tau = t$ we obtain such an expression.

10.2.1 Exercises and comments

(1) Note that, in the case of a noisy plant equation, it is only because of the RSCEP that we can regard the plant equation as a constraint between observables, which can be coped with by Lagrangian methods. The plant equation (10.15) at a future moment in time is a relation between unobservables, and to regard it as a 'constraint' would be an empty notion. However, the RSCEP gives us a criterion for estimating unobservables, and (10.15) then indeed appears as a constraint between these estimates.

(2) The risk-neutral case $\theta = 0$ is indeed degenerate, in that relation (10.17) is no longer clearly meaningful. As θ tends to zero then $\varepsilon_\tau^{(t)}$ also tends to zero for $\tau > t$, so that $x_\tau^{(t)}$ (then identifiable with the ML predictor of x_τ) is identical with the value of x_τ for the noise-free process. In this limit $\lambda_\tau^{(t)}$ still makes sense as the multiplier associated with the constraint (10.10) of the noise-free plant equation at τ, which means that, although $\varepsilon_\tau^{(t)}$ has a zero limit, $\varepsilon_\tau^{(t)}/\theta$ has a non-zero limit. See also Exercise 10.3.2.

10.3 THE PATH INTEGRAL AND THE HAMILTONIAN EQUATIONS FOR THE CASE OF IMPERFECT OBSERVATION

If we consider the general time-homogeneous linear model with imperfect observation, formulated in Section 10.1, then the expression for the total stress is

$$\mathbb{S} = \sum_{\tau=0}^{h-1} c_\tau + \mathbb{C}_h + \theta^{-1}\left[\mathbb{D}_0 + \sum_{\tau=1}^{h} d_\tau\right] \quad (10.20)$$

Here c_τ has the evaluation (10.7) and d_τ the evaluation

$$d_\tau = \left[\begin{bmatrix}\varepsilon\\\eta\end{bmatrix}'\begin{bmatrix}N & L\\L' & M\end{bmatrix}^{-1}\begin{bmatrix}\varepsilon\\\eta\end{bmatrix}\right]_\tau \quad (10.21)$$

with the understanding that ε and η are to be expressed in terms of x, y, u and α by substitution from the plant and observation relations, (10.1) and (10.2), at time τ.

As in the last section, we can use Lagrange multiplier vectors λ_τ and μ_τ to take account of these relations, so forming the Lagrangian form

$$\mathbb{S} + 2\sum_{\tau=1}^{h}[\lambda'(\mathscr{A}x + \mathscr{B}u - \alpha - \varepsilon) + \mu'(y + \mathscr{C}x - \eta)]_\tau$$

Extremizing with respect to the noise variables we obtain the relationship

$$\theta\begin{bmatrix}N & L\\L' & M\end{bmatrix}\begin{bmatrix}\lambda\\\mu\end{bmatrix}_\tau = \begin{bmatrix}\varepsilon\\\eta\end{bmatrix}_\tau \quad (10.22)$$

10.3 THE PATH INTEGRAL AND THE HAMILTONIAN EQUATIONS

and are left with the expression

$$\mathbb{I}(x,u,\lambda,\mu) = \theta^{-1}\mathbb{D}_0 + \sum_{\tau=1}^{h} [2\lambda'(\mathscr{A}x + \mathscr{B}u - \alpha) + 2\mu'(y + \mathscr{C}x) - \theta v]_\tau$$
$$+ \sum_{\tau=0}^{h-1} c_\tau + \mathbb{C}_h \tag{10.23}$$

where

$$v_\tau = v(\lambda_\tau, \mu_\tau) = \left[\begin{bmatrix} \lambda \\ \mu \end{bmatrix}' \begin{bmatrix} N & L \\ L' & M \end{bmatrix} \begin{bmatrix} \lambda \\ \mu \end{bmatrix} \right]_\tau \tag{10.24}$$

We shall regard expression (10.23) as our path integral. The effect of the $(\varepsilon, \eta) \to (\lambda, \mu)$ transformation has been to clear the expression for the stress of matrix inverses. Also, we know from Chapter 5 that it is in terms of the (λ, μ) variables that we see the 'past/estimation' problem as exactly dual to the 'future/control' problem with its associated variables (x, u).

As before, let t denote the current instant, so that one possesses the information on which the choice of u_t is to be based. According to the RSCEP we determine the optimal value of u_t by seeking the extreme of $\mathbb{I}$ with respect to decisions currently unmade (u_τ for $t \leq \tau < h$), quantities currently unobservable (x_τ for $0 \leq \tau \leq h$ and y_τ for $t < \tau \leq h$) and now also with respect to the dual variables λ_τ and μ_τ for $0 < \tau \leq h$. The nature of the extreme is now so manifold that we summarize the situation in Table 10.3.1.

Table 10.3.1. The nature of the extreme of the path integral $\mathbb{I}(x, u, \lambda, \mu)$ with respect to its arguments

	$\theta > 0$	$\theta < 0$
u	min	min
x, y	min	max
λ, μ	max	min

In fact, we shall simply write down the stationarity conditions for $\mathbb{I}$ with respect to these variables. The conditions that $\mathbb{S}$ (and so $\mathbb{I}$) should have as only stationary point a saddle point appropriate for the extremum required by Table 10.3.1 is equivalent to the condition that θ should have a value such that the minimization of $\gamma_\pi(\theta)$ with respect to π is a properly posed problem.

We gain some formal insight by writing down the stationarity conditions for $\mathbb{I}$ with respect to *all* variables x, u, λ and μ at a given time τ. In fact, if $\tau < t$ then u_τ is known, and if $\tau > t$ then y_τ is unobserved, which implies effectively that $\mu_\tau = 0$ (see Theorem 10.3.1 and Exercise 10.3.1). Nevertheless, the full set of stationarity conditions would be just

$$\begin{bmatrix} R & S' & \bar{\mathscr{A}} & \bar{\mathscr{C}} \\ S & Q & \bar{\mathscr{B}} & 0 \\ \mathscr{A} & \mathscr{B} & -\theta N & -\theta L \\ \mathscr{C} & 0 & -\theta L' & -\theta M \end{bmatrix} \begin{bmatrix} x \\ u \\ \lambda \\ \mu \end{bmatrix}_\tau = \begin{bmatrix} R\bar{x} + S'\bar{u} \\ S\bar{x} + Q\bar{u} \\ \alpha \\ -y \end{bmatrix}_\tau \quad (10.25)$$

The matrix of operators on the left has a pleasingly symmetric and partitioned form, which fully expresses the duality of control and estimation. In fact, of course, system (10.25) holds only in the modified form that, for $\tau < t$, the second equation is replaced by the specification of u_τ, and that, for $\tau > t$, the final equation is replaced by the specification $\mu_\tau = 0$. Amplifying this argument, we deduce:

Theorem 10.3.1. *Assume that θ has a value such that the control-optimization problem is well posed. Then the stationarity conditions for the path integral $\mathbb{I}(x, u, \lambda, \mu)$ for given W_t (which determine the optimal u_t in closed-loop form) yield the pair of equation systems*

$$\begin{bmatrix} R & S' & \bar{\mathscr{A}} \\ S & Q & \bar{\mathscr{B}} \\ \mathscr{A} & \mathscr{B} & -\theta N \end{bmatrix} \begin{bmatrix} x \\ u \\ \lambda \end{bmatrix}^{(t)}_\tau = \begin{bmatrix} R\bar{x} + S'\bar{u} \\ S\bar{x} + Q\bar{u} \\ \alpha + \theta L\mu \end{bmatrix}^{(t)}_\tau \quad (t \leq \tau < h) \quad (10.26)$$

$$\begin{bmatrix} N & L & \mathscr{A} \\ L' & M & \mathscr{C} \\ \bar{\mathscr{A}} & \bar{\mathscr{C}} & -\theta R \end{bmatrix} \begin{bmatrix} -\theta\lambda \\ -\theta\mu \\ x \end{bmatrix}^{(t)}_\tau = \begin{bmatrix} \alpha - \mathscr{B}u \\ -y \\ \theta S'u \end{bmatrix}^{(t)}_\tau \quad (0 < \tau \leq t) \quad (10.27)$$

where $\mathcal{T}$ operates on the subscript τ. These are subject to the subsidiary conditions

$$u^{(t)}_\tau \text{ prescribed } (\tau < t)$$
$$\mu^{(t)}_\tau = 0 \quad (\tau > t) \quad (10.28)$$

and the terminal conditions

$$(\lambda, \mu)_\tau = 0 \quad (\tau \leq 0; \tau > h) \quad (10.29)$$

$$\frac{1}{\theta} \frac{\partial \mathbb{D}_0}{\partial x_\tau} + (\bar{\mathscr{A}}\lambda + \bar{\mathscr{C}}\mu)_\tau = 0 \quad (\tau \leq 0) \quad (10.30)$$

$$\frac{\partial \mathbb{C}_h}{\partial x_h} + \lambda_h = 0 \quad (10.31)$$

One can make the identification

$$\theta \begin{bmatrix} N & L \\ L' & M \end{bmatrix} \begin{bmatrix} \lambda \\ \mu \end{bmatrix}^{(t)}_\tau = \begin{bmatrix} \varepsilon \\ \eta \end{bmatrix}^{(t)}_\tau \quad (10.32)$$

10.3 THE PATH INTEGRAL AND THE HAMILTONIAN EQUATIONS

We have now introduced the t-superscript to emphasize that optimization is taking place from time t. However, in relations (10.29)–(10.31) we have omitted the superscript for typographical simplicity. We have taken as a convention (in which we are not always consistent) that $\partial \mathbb{D}_0/\partial x_h$ is the *column* vector of derivatives of $\mathbb{D}_0$ with respect to the components of x_h, etc.

Proof. This is very much a matter of verification. Note first that, since there is no observation relation (10.2) for $\tau > t$ we effectively have $\mu_\tau^{(t)} = 0$ in this range. For an approach which retains these constraints and estimates future observations, see Exercise 10.3.1.

We thus deduce equations (10.28), and systems (10.26) and (10.27) are just the restrictions of system (10.25) which will apply in the two ranges.

It is necessary to write down relation (10.29), because λ_τ or μ_τ for τ not in $[l, h]$ may occur in some of the equations of (10.26) or (10.27), but simply do not appear in the path integral (10.23), and so are effectively zero. Conditions (10.30) and (10.31) follow from x_τ extremes for which either $\mathbb{D}_0$ or $\mathbb{C}_h$ make a contribution. Relation (10.32) follows from (10.22). □

The appreciation of this rather concentrated characterization and the unravelling of its implications take some time. The systems (10.26) and (10.27) are what we regard as our 'Hamiltonian equations'. Note that system (10.26) is just the system (10.12) derived in the perfect-observation case of the previous section, except that it is now coupled to system (10.27) by the occurrence of μ_τ in the right-hand member (having an effect at $\tau = t$) and by the fact that past process values x_τ ($\tau \leq t$) are now also estimated.

We have already commented upon system (10.26) in the previous section, noting that it can be regarded as a statement of the maximum principle which incorporates risk sensitivity so neatly that it makes this appear as the element which the theory has been missing, and which makes transference of the principle to the stochastic case possible. The system (10.27) is obviously the complete analogue: the dual of (10.26) in the same sense that past and future are mutually dual and control and estimation are mutually dual.

The matrices of operators occurring in (10.26) and (10.27) are certainly fundamental; we shall denote them by

$$\Phi(\mathcal{T}) = \begin{bmatrix} R & S' & \bar{\mathcal{A}} \\ S & Q & \bar{\mathcal{B}} \\ \mathcal{A} & \mathcal{B} & -\theta N \end{bmatrix} \tag{10.33}$$

$$\Psi(\mathcal{T}) = \begin{bmatrix} N & L & \mathcal{A} \\ L' & M & \mathcal{C} \\ \bar{\mathcal{A}} & \bar{\mathcal{C}} & -\theta R \end{bmatrix} \tag{10.34}$$

We shall examine the first implications of this structure in Section 10.4, consider the structure divorced from the particular control context in Section 10.5, and set out directions for further analysis in Section 10.6.

10.3.1 Exercises and comments

(1) Assume that the terms accounting for future observation constraints are retained in the path integral (10.23). Extremization with respect to y_τ for $\tau > t$ then implies that μ_τ must be zero if this extreme is to be finite. Relation (10.32) then reduces to (10.17) plus an estimate $\theta L \lambda_\tau^{(t)} = L N^{-1} \varepsilon_\tau^{(t)}$ of η_τ. Extremization with respect to μ_τ for $\tau > t$ then yields the predictive relation

$$y_\tau^{(t)} = C x_\tau^{(t)} + \theta L \lambda_\tau^{(t)} = C x_\tau^{(t)} + \eta_\tau^{(t)}$$

(2) As we converge to the risk-neutral case by letting θ tend to zero then, as was observed in Exercise 10.2.2, $\lambda_\tau^{(t)}$ has a limit value (and $\mu_\tau^{(t)}$ we know to be zero) for $\tau > t$. We may suspect from the form of (10.27), however, that $\theta \lambda_\tau^{(t)}$ and $\theta \mu_\tau^{(t)}$ have non-zero limit values for $\tau \leq t$, indeed related to estimates of *past* noise values by (10.32). These are the values which we denoted by l_τ and m_τ in Chapter 5.

(3) The missing entries in the matrix of system (10.25) would be filled if we modified the observation relation (10.2) to $y + \mathscr{C}x + \mathscr{D}u = \eta$. However, this adds no real generality, because u_t is as yet undetermined when y_t is observed. Thus we must have $D_0 = 0$, and so $\mathscr{D}u_t$ is a known quantity before y_t is observed.

(4) The form of equation (10.25) invites us to define 'supervariables'

$$\mathfrak{x} = \begin{bmatrix} x \\ u \end{bmatrix} \text{ and } \lambda = \begin{bmatrix} \lambda \\ \mu \end{bmatrix}$$

and rewrite (10.25) as

$$\begin{bmatrix} \mathfrak{R} & \bar{\mathfrak{A}} \\ \mathfrak{A} & -\theta \mathfrak{N} \end{bmatrix} \begin{bmatrix} \mathfrak{x} \\ \lambda \end{bmatrix} = \cdots \quad (10.25a)$$

in terms of corresponding 'supermatrices' $\mathfrak{A}$, $\mathfrak{R}$ and $\mathfrak{N}$. One would interpret $\mathfrak{x}$ as the *system variable*, λ as its conjugate, and $\mathfrak{A}$, $\mathfrak{R}$ and $\mathfrak{N}$ as pertaining to system dynamics, costs and noise statistics, respectively. Thanks to the RSCEP, the conjugate variable λ has a meaning even in this stochastic case, either as $\partial \bar{\mathbb{1}}/\partial \mathfrak{x}$ (c.f. Exercise 5.2.1) or as $(\theta \mathfrak{N})^{-1} \varepsilon^{(t)}$ (c.f. (10.32)).

We shall use this system formulation occasionally for its helpful compactness; see e.g. Section 10.7 and Chapter 17. However, we cannot move wholly into this formulation because, as emphasized after equation (10.25), optimization is subject to the constraints that past u and future μ are known. This, of course, is why the single-system (10.25) must be replaced by the two coupled systems (10.26) and (10.27).

(5) The reductions of Exercises 3.7.4 and 3.7.5 can be generalized. Consider

the system

$$\Phi(\mathcal{T})\zeta = 0$$

where Φ is defined by (10.33) and ζ has the components x, u and λ. If $Q > 0$ then we can eliminate u to obtain the system

$$\Phi_1(\mathcal{T})\begin{bmatrix} x \\ \lambda \end{bmatrix} = 0$$

where

$$\Phi_1 = \begin{bmatrix} R & \bar{\mathcal{A}} \\ \mathcal{A} & -\mathcal{J} \end{bmatrix} \quad (10.35)$$

Here R and $\mathcal{A}$ should in fact be given the values $R - S'Q^{-1}S$ and $\mathcal{A} - \mathcal{B}Q^{-1}S$ obtained under the normalization of S to zero, and $\mathcal{J} = \mathcal{B}Q^{-1}\bar{\mathcal{B}} + \theta N$. If R were positive definite then we could also eliminate x, to obtain

$$\Phi_2(\mathcal{T})\lambda = 0$$

where

$$\Phi_2 = \mathcal{A}R^{-1}\bar{\mathcal{A}} + \mathcal{J} \quad (10.36)$$

(R and $\mathcal{A}$ being again the normalized values).

10.4 THE POINT AND ADVANTAGES OF THE PATH-INTEGRAL APPROACH

We shall speak of the approach as the 'path-integral approach', since the RSCEP does indeed reduce the optimization problem to the free extremization of a quadratic path integral. We shall speak of the two equations systems (10.26) and (10.27) as the 'Hamiltonian equations', for reasons explained in Appendix 9. These are the forms that the stationarity conditions for the path integral take when we make the time-homogeneity assumptions of Section 10.1 and allow for the constraints of the plant and observation relations by Lagrangian multipliers λ and μ.

The first point to note is the elegance of the result: that the stationarity conditions on the path integral reduce to the compact equation systems (10.26) and (10.27) with the operators Φ and Ψ Hermitian in that $\bar{\Phi} = \Phi$ and $\bar{\Psi} = \Psi$. This formalism copes naturally with the introduction of higher-order operators $\mathcal{A}$, $\mathcal{B}$ and $\mathcal{C}$ into the plant and observation relations, and indeed itself suggests further generalizations which are now formally immediate (see Section 10.6).

The complete duality of past and future is latent in this subject, and manifest if the right formalism is used, as is evident from comparison of the two equation systems.

The second point, which we have already made in Sections 10.2 and 10.3 but which will bear repetition, is the natural incorporation of risk sensitivity. In Section 10.2 we began with the rather obvious path-integral characterization of the deterministic problem. This generalized to the stochastic case by the simple insertion of the noise-power term $-\theta N$ into the empty entry of $\Phi(\mathcal{T})$. The risk-neutral 'maximum principle' for the estimation problem would be the maximum likelihood principle. This requires transformation to conjugate variables to bring it into completely dual form. Risk sensitivity is then incorporated by the simple insertion of the cost-pressure term $-\theta R$ into the empty entry of $\Psi(\mathcal{T})$.

One might now make the following qualitative assertion.

Theorem 10.4.1. *The risk-sensitive certainty-equivalence principle, with plant and observation constraints accounted for by Lagrange multipliers, yields the natural statement of a stochastic maximum principle, in that equations (10.26) and (10.27) provide the statement of just such a principle, with all variables calculable from quantities observable at the current instant.*

'*Proof*'. We use inverted commas here because the term 'natural statement' is not objectively defined, and so both assertion and proof are, to that degree, qualitative. The maximum principle is conventionally stated for a first-order continuous-time control problem. However, in the present case it would amount to the analysis set out in equations (10.10)–(10.14), with the interpretation of λ_τ as the Lagrange multiplier associated with the constraint supplied by the plant equation at time τ.

The RSCEP, applied in the case of an additively decomposable cost, has obviously supplied us with a generalized version of this principle. The additional set of multipliers μ_τ is needed if the process variable is only imperfectly observable, but all quantities $(x, u, \lambda, \mu, \varepsilon, \eta)_\tau^{(t)}$ occurring are well defined, as calculable from the information available at time t. □

There have been many attempts to produce a 'stochastic maximum principle' which generalizes the extremal principle associated with the path integral (10.11). However, they have involved notions such as 'future stimulating prices', which are essentially Lagrange multipliers, random in that they are associated with constraints between quantities not currently observable. Such principles are not operational; one ends up with equations not soluble in terms of current observables. In our case the multipliers are associated with constraints between estimates, and all quantities are calculable in terms of current observables.

Interestingly, it is only by adopting the risk-sensitive formulation that one acquires the perspective needed to reveal the right formalism. It is for this reason that the risk-neutral case appears as, indeed, degenerate in some sense.

The last point is that already touched upon in Sections 3.7 and 5.4. This is that, in the case of an infinite horizon and an infinite history, one will expect canonical factorizations of Φ and Ψ to reduce equation systems (10.26) and

10.4 THE POINT AND ADVANTAGES OF THE PATH-INTEGRAL APPROACH

(10.27) so that they give virtually an immediate solution for the optimal control. To repeat the formal argument, suppose horizon and history infinite, so that (10.26) and (10.27) hold for all $\tau \geq t$ and all $\tau \leq t$, respectively. Write these systems as

$$\Phi \zeta_\tau^{(t)} = \omega_\tau \qquad (\tau \geq t) \qquad (10.37)$$

$$\Psi \chi_\tau^{(t)} = \rho_\tau \qquad (\tau \leq t) \qquad (10.38)$$

Assume canonical factorizations

$$\Phi = \Phi_- \Phi_0 \Phi_+ \qquad (10.39)$$

$$\Psi = \Psi_+ \Psi_0 \Psi_- \qquad (10.40)$$

where matrices with a zero subscript are constant (i.e. independent of $\mathcal{T}$), those with a plus subscript are such that they and their inverses properly operate into the past, and those with a minus subscript are such that they and their inverses properly operate into the future. Under appropriate regularity conditions equations (10.37) and (10.38) will then be partially invertible to

$$\Phi_+ \zeta_\tau^{(t)} = \Phi_0^{-1} \Phi_-^{-1} \omega_\tau \qquad (\tau \geq t) \qquad (10.41)$$

$$\Psi_- \chi_\tau^{(t)} = \Psi_0^{-1} \Psi_+^{-1} \rho_\tau \qquad (\tau \leq t) \qquad (10.42)$$

In the risk-neutral Markov case the vectors ω and ρ are known, and equations (10.41) and (10.42) for $\tau = t$ supply expressions for the optimal control and the estimate of state. The two systems are indeed linked only by the fact that the estimate $x_t^{(t)}$ derived from equation (10.42) at t must be substituted into equation (10.41) at t, where it occurs.

In the higher-order and risk-sensitive cases there is a more extensive and reciprocal coupling, but it is still true that the partial inversions (10.41) and (10.42) reduce the two semi-infinite equation systems (10.37) and (10.38) to the solution of something like np scalar matching conditions. We return to these matters in Chapter 14.

10.4.1 Exercises and comments

(1) As in Section 5.2, we could rewrite the partial inversion (10.42) as the two-stage recursion (5.32) and (5.33), the latter constituting the general version of the Kalman filter. In the risk-sensitive case there are coupling terms which must be taken into account, but it is still true that relation (5.33) is to be regarded as the general statement of the Kalman filter.

(2) The same considerations hold for the forward system (10.37). We can write this in the two-stage form

$$\Phi_- \sigma_\tau = \omega_\tau \qquad (10.43)$$

$$\Phi_0 \Phi_+ \zeta_\tau^{(t)} = \sigma_\tau \qquad (10.44)$$

In the homogeneous case $\omega = 0$ the optimal control rule would be determined by

$$\Phi_+ \zeta_\tau^{(t)} = 0 \qquad (10.45)$$

Relation (10.45) generates the feedback component of optimal control (in the perfect-observation case); relation (10.43) generates the feedforward component. We have already commented on the fact that the relations generating these components are mutually conjugate (e.g. after equation (3.69)).

10.5 GENERAL PATH-INTEGRAL FORMALISM

Some of the basic structure of a path-integral extremization was obscured in Sections 10.2 and 10.3 by the details of our particular problem. For this reason it is actually simplifying to consider such problems in general form.

We stay with a discrete-time formulation for the moment, so that our 'integral' is in fact a sum. Assume that one has the quadratic path integral in a vector variable ζ_t:

$$\mathbb{I}(\zeta) = \sum_\tau \left[\left(\sum_j \zeta_\tau' G_j \zeta_{\tau-j} \right) - \omega_\tau' \zeta_\tau - \zeta_\tau' \omega_\tau \right] + \text{end effects} \qquad (10.46)$$

Here the term 'end effects' represents special contributions to $\mathbb{I}$ at initial and final τ. The matrix and vector coefficients G_j and ω_τ are known, and we assume the form Hermitian in that

$$G_{-j} = G_j' \qquad (10.47)$$

for all relevant j. This is indeed just a matter of normalization.

We assume the problem one of *forward optimization* in that ζ_τ is supposed *known* for $\tau \le t$ and is to be chosen *freely* for $\tau > t$ in such a way as to render the form stationary. Here t is a prescribed moment in time, the current instant, and, by speaking of 'stationarity' we leave open the possibility that we may be minimizing with respect to some components of ζ and maximizing with respect to others.

We have thus formulated a general version of the forward-optimization problem of Section 10.2, for which ζ would be the column vector with components x, u and λ. Any general conclusions will have an immediate 'backward' analogue, and presumably be useful for the treatment of coupled forward and backward problems.

The stationarity condition with respect to ζ_τ is then

$$\Phi(\mathcal{T}) \zeta_\tau^{(t)} = \omega_\tau \qquad (t < \tau < h) \qquad (10.48)$$

where h is the horizon point and

$$\Phi(\mathcal{T}) = \sum_j G_j \mathcal{T}^j \qquad (10.49)$$

Equation (10.48) is subject to boundary conditions: prescription of ζ_τ for $\tau \le t$ and special terminal conditions at $\tau = h$.

It is formalism as simple as this which lies behind the derivation of equations (10.26), with the matrices G_j determined by the specification of costs and stochastic dynamics for the problem. In much subsequent work we shall find it simpler to use this general formulation, reverting to the more special formulation of the control problem only when we wish for final results or to appeal to a more special structure.

10.6 QUESTIONS TO BE RESOLVED

In this chapter we have tried to set out the formalism so that the general pattern and possibilities can be seen. The analysis has been rigorous (at least in the discrete-time finite-horizon case). Theorem 10.3.1 is subject only to the proviso that θ be not so negative that $\mathbb{I}$ loses its required saddle-point character. However, a great many points remain before one has a clear picture and a workable technique, and it is to these that we shall now largely devote ourselves. It is best to list the unresolved points.

(1) One must determine conditions which ensure that the canonical factorizations (10.39) and (10.40) exist. One would wish also to determine the nature of the factors as closely as possible, perhaps under some natural normalization.

(2) One should develop techniques for numerical determination of the factors in given cases. Similarly, whatever further calculations are needed to fully determine the optimal control from the reduced equations (10.41) and (10.42) should be clarified.

(3) One should deduce the higher-order versions of the Riccati equations for Π and V and relate solution of these to the achievement of a canonical factorization.

(4) The dependence of the problem on the risk-sensitivity parameter θ should be clarified. For example, it seems that, as θ tends to zero, then $\lambda_\tau^{(t)}$ tends to a non-zero limit and $\theta \lambda_\tau^{(t)}$ tends to zero for $\tau > t$, but that $\theta \lambda_\tau^{(t)}$ and $\theta \mu_\tau^{(t)}$ tend to the limits l_τ and m_τ of Section 5.3 for $\tau \le t$.

(5) The form of the operators $\Phi(\mathcal{T})$ and $\Psi(\mathcal{T})$ makes clear that further generalization is almost immediate. For example, one could generalize Φ to

$$\Phi(z) = \begin{bmatrix} R(z) & \overline{S(z)} & \overline{\mathcal{A}(z)} \\ S(z) & Q(z) & \overline{\mathcal{B}(z)} \\ \mathcal{A}(z) & \mathcal{B}(z) & -\theta N(z) \end{bmatrix} \qquad (10.50)$$

where

$$R(z) = \sum_j R_j z^j \qquad (10.51)$$

etc. This would correspond to the generalization of a cost component such as $x'_\tau R x_\tau$ to

$$\sum_j x'_\tau R_j x_{\tau-j} \tag{10.52}$$

and to the generalization of the Gaussian white-noise process $\{\varepsilon_t\}$ to a stationary Gaussian process with autocovariance matrix

$$\text{cov}(\varepsilon_t, \varepsilon_{t-j}) = N_j \tag{10.53}$$

That is, it seems as though the formulation would also cope with autocorrelation in the noise and cross-products (over time) in the cost function. Our previously listed queries would then have to be answered also in this more general case; we consider the matter to some extent in Chapters 13 and 16.

10.7 SYSTEM NOTATION

We noted in Exercise 10.3.4 that the form of equations (10.25) invited us write these equations in the condensed form (10.25a). However, the fact that equations (10.25) must in fact be replaced by the two sets (10.26) and (10.27) means that we cannot condense notation to this degree.

However, in discussing the 'forward' system (10.26) we shall indeed sometimes write this as

$$\begin{bmatrix} \mathfrak{R} & \bar{\mathfrak{A}} \\ \mathfrak{A} & -\theta N \end{bmatrix} \begin{bmatrix} \mathfrak{x} \\ \lambda \end{bmatrix} = \cdots \tag{10.54}$$

where then

$$\mathfrak{x} = \begin{bmatrix} x \\ u \end{bmatrix} \quad \mathfrak{A} = [\mathscr{A} \; \mathscr{B}] \quad \mathfrak{R} = \begin{bmatrix} R & S' \\ S & Q \end{bmatrix}$$

We shall refer to this as the *system* (rather than the *process*) notation. This is a convention which will be particularly useful in Section 13.2 and Chapter 17, when it greatly simplifies the calculations. The view that it implies is a valuable one. For example, the plant equation becomes

$$\mathfrak{A}\mathfrak{x} = \alpha + \varepsilon \tag{10.55}$$

and the view of control optimization, that u is to be chosen optimally, becomes rather that $\mathfrak{x}$ is to be chosen optimally subject to the constraint (10.55).

10.8 NOTES ON THE LITERATURE

The Pontryagin maximum principle, of course, runs right through the literature, and the Hamiltonian characterization of the conditions it gives rise to in the LQ case are found recurrently; see, particularly, Graves and Telser (1967) and Telser

10.8 NOTES ON THE LITERATURE

and Graves (1968, 1972). The corresponding ideas in the estimation context are much less developed, but indications of them are found in, for example, Ljung *et al.* (1976), Friedlander *et al.* (1976, 1977), van Dooren (1981) and Adams *et al.* (1984). General Hamiltonian formulations in both contexts are given explicitly in the supplementary chapters of Whittle (1983b).

A paper by Hagander (1973) is notable in that it explicitly formulates the optimization of both control and estimation (in the Markov case) as a factorization problem. Treating a special case as it does, it does not reveal the full formalism, but the paper undoubtedly presages the approach taken in this book.

None of these formulations consider risk sensitivity. The material of Section 10.3, indicating how a path-integral formulation could incorporate risk sensitivity and imply an operational stochastic maximum principle, was presented in Whittle and Kuhn (1986).

CHAPTER 11

The Markov case: recursions and factorizations

Despite the fact that we wish to understand the general structure of the higher-order case, it is better, as always, to begin with the first-order (Markov) model. This is the case for which results are most transparent and explicit, and all others can be reduced to it (although at a cost, see Exercise 1.3.2). It is especially interesting to take up the Markov case in the general formalism of Section 10.5.

11.1 FORMALISM IN THE MARKOV CASE

In terms of the general path-integral formalism of Section 10.5 the Markov case is distinguished by

$$G_j = 0 \quad (|j| > 1) \tag{11.1}$$

It is actually better to pursue the formalism in this general setting for a while before considering the implications of the particular forms that the matrices G_{-1}, G_0 and G_1 have in the control problem. As in Section 10.5, we consider a forward-optimization problem (for $\tau > t$) for definiteness. The analysis of a backward optimization then follows by simple time reversal.

In extremizing a path-integral quadratic in ζ (such as (10.46)) we shall have occasion to introduce analogues of all the quantities F, Π and Γ, etc. of the LQ control problem originally formulated in Chapter 3. In an attempt both to economize on notation without confusion and to point the analogy we shall use the same notation in bold type, e.g.: $\boldsymbol{F}, \boldsymbol{\Pi}, \boldsymbol{\Gamma}, \ldots$.

Let us consider the recursion first. As for the original control problem, we can define a value function by considering extremization of the partial integral (10.46) from a general initial co-ordinate ζ_t at time t:

$$\boldsymbol{F}_t(\zeta_t) = \underset{\zeta_\tau : \tau > t}{\text{stat}} \left[\sum_{\tau=t}^{h-1} \mathbf{c}_\tau + \text{terminal cost} \right] \tag{11.2}$$

Here

$$\mathbf{c}_\tau = \zeta'_\tau G_0 \zeta_\tau + \zeta'_\tau G_{-1} \zeta_{\tau+1} + \zeta'_{\tau+1} G_1 \zeta_\tau - \omega'_\tau \zeta_\tau - \zeta'_\tau \omega_\tau \tag{11.3}$$

in accordance with the Markov assumption (11.1). Assume that the terminal cost takes the quadratic form

$$F_h(\zeta_h) = [\zeta'\Pi\zeta - 2\sigma'\zeta + \gamma]_h \tag{11.4}$$

The extremization in (11.2) will, in general, be a mixture of maximization and minimization operations. For the moment we shall assume that we shall be led to the correct extreme if we simply seek a stationary point, as indicated. This view is legitimized for the control case in the next section.

The value function F then obviously obeys the optimality equation

$$F_t(\zeta_t) = \operatorname*{stat}_{\zeta_{t+1}} [\mathbf{c}_t + F_{t+1}(\zeta_{t+1})] \qquad (t < h) \tag{11.5}$$

whence we derive

Theorem 11.1.1. *The value function defined by* 11.2 *has the quadratic evaluation*

$$F_t(\zeta_t) = [\zeta'\Pi\zeta - 2\sigma'\zeta + \gamma]_t \qquad (t \leq h) \tag{11.6}$$

Here Π, σ *and* γ *obey the backward recursions*

$$\Pi_t = G_0 - G_{-1}\Pi_{t+1}^{-1}G_1 \tag{11.7}$$

$$\sigma_t = \omega_t - G_{-1}\Pi_{t+1}^{-1}\sigma_{t+1} \tag{11.8}$$

$$\gamma_t = \gamma_{t+1} - \sigma'_{t+1}\Pi_{t+1}^{-1}\sigma_{t+1} \qquad (t < h) \tag{11.9}$$

The extremizing ζ_{t+1} *is determined by*

$$\Pi_{t+1}\zeta_{t+1} + G_1\zeta_t = \sigma_{t+1} \tag{11.10}$$

These results all follow by conventional calculations; relation (11.7) is the Riccati recursion and (11.10) the 'optimally controlled plant equation'. What is striking is how much more simple are recursions (11.7)–(11.9) in this general formulation than for the specific control problem (see e.g. Sections 3.2 and 3.3).

Relations (11.8) and (11.10) could be rewritten

$$\sigma_t = \omega_t + \Gamma'_t\sigma_{t+1} \tag{11.11}$$

$$\zeta_{t+1} = \Gamma_t\zeta_t + \Pi_{t+1}^{-1}\sigma_{t+1} \tag{11.12}$$

in terms of the gain matrix

$$\Gamma_t = -\Pi_{t+1}^{-1}G_1 \tag{11.13}$$

We would expect that, under suitable regularity conditions, the matrix Π_t would tend to a limit Π as the time-to-go $h - t$ became infinite, and that the corresponding limit Γ of Γ_t would be a stability matrix. The relation of these limits to the canonical factorization of

$$\Phi(z) = G_{-1}z^{-1} + G_0 + G_1z \tag{11.14}$$

is immediate.

11.2 THE RELATION BETWEEN THE GENERAL AND SPECIAL PROBLEMS

Theorem 11.1.2. *If the limit Π exists then*
$$\Phi(z) = (\Pi + G_{-1}z^{-1})\Pi^{-1}(\Pi + G_1 z) \tag{11.15}$$
is a canonical factorization of Φ. This can be alternatively expressed
$$\Phi(z) = \overline{(I - \Gamma z)}\Pi(I - \Gamma z) \tag{11.16}$$

Proof. The coefficient of z in the expansion of expression (11.15) is indeed G_1. Equating the absolute term to G_0 we see that we require the relation
$$G_0 = \Pi + G_{-1}\Pi^{-1}G_1 \tag{11.17}$$
which is indeed just the equilibrium form of the Riccati equation (11.7).

Expression (11.15) thus constitutes a valid factorization of Φ, equivalent to (11.16) by (11.13). It will be canonical if Γ is a stability matrix, which we expect and shall establish. □

A slight rewording of this last assertion is to say that $\Pi + G_1 z$ must be a canonical factor because we expect relation (11.10) to be a stable forward recursion for ζ.

11.2 THE RELATION BETWEEN THE GENERAL AND SPECIAL PROBLEMS

At the cost of losing the simplicity of the general formulation and the momentum of analysing it, we stop to establish the relation between the general and special problems. This has merit in that the convergence of Π_t to a limit with increasing time-to-go can be deduced from that of Π_t, which we know we can establish under appropriate conditions (Theorem 9.2.1).

If we consider the optimal control problem in the case $\tau > t$ then essentially we are considering solution of the F-recursion. This would amount to the determination of the optimal control with perfect state observation.

The particular Φ we are concerned with is then
$$\Phi(z) = \begin{bmatrix} R & S' & \bar{\mathcal{A}} \\ S & Q & \bar{\mathcal{B}} \\ \mathcal{A} & \mathcal{B} & -\theta N \end{bmatrix} \tag{11.18}$$

which in the Markov case reduces to
$$\Phi(z) = \begin{bmatrix} R & S' & I - A'z^{-1} \\ S & Q & -B'z^{-1} \\ I - Az & -Bz & -\theta N \end{bmatrix} \tag{11.19}$$

corresponding to

$$G_0 = \begin{bmatrix} R & S' & I \\ S & Q & 0 \\ I & 0 & -\theta N \end{bmatrix} \quad (11.20)$$

$$G_1 = G'_{-1} = -\begin{bmatrix} 0 & 0 & 0 \\ 0 & 0 & 0 \\ A & B & 0 \end{bmatrix} \quad (11.21)$$

In discussing questions such as horizon stability and canonical factorization we can revert to the homogeneous case, with quantities such as ω and $\boldsymbol{\sigma}$ (or α, $\bar{x}$, $\bar{u}$ and σ) all taken as zero. This is because these questions depend wholly on the behaviour of the matrices Π_t (or $\tilde{\Pi}_t$) with increasing horizon.

In the control context we identify ζ with the column vector with components x, u and λ. With the G-matrices having the particular evaluations (11.20) and (11.21) definition (11.2) then becomes

$$F_t(\zeta_t) = \underset{\zeta_\tau: \tau > t}{\text{stat}} \left[\sum_{\tau=t}^{h-1} [c(x_\tau, u_\tau) + 2\lambda'_\tau x_\tau - 2\lambda'_{\tau+1}(Ax_\tau + Bu_\tau) - \theta\lambda'_\tau N\lambda_\tau] + (x'\Pi x)_h \right]$$

$$= c(x_t, u_t) + 2\lambda'_t x_t - \theta\lambda'_t N\lambda_t + \underset{\varepsilon_{t+1}}{\text{ext}} [F_{t+1}(Ax_t + Bu_t + \varepsilon_{t+1}) - \theta^{-1}\varepsilon'_{t+1} N^{-1}\varepsilon_{t+1}]$$

(11.22)

Now, we know that $F_{t+1}(x) = x'\Pi_{t+1}x$. It follows then that (11.22) reduces to

$$F_t(\zeta) = c(x, u) + 2\lambda'x - \theta\lambda'N\lambda + (Ax + Bu)'\tilde{\Pi}_{t+1}(Ax + Bu) \quad (11.23)$$

where

$$\tilde{\Pi}_{t+1} = (\Pi_{t+1}^{-1} + \theta N)^{-1} \quad (11.24)$$

(c.f. equation (7.21)). We thus deduce

Theorem 11.2.1. *The dependence of Π upon $\tilde{\Pi}$ is expressed by*

$$\Pi_t = \begin{bmatrix} R + A'\tilde{\Pi}_{t+1}A & S' + A'\tilde{\Pi}_{t+1} & I \\ S + B'\tilde{\Pi}_{t+1}A & Q + B'\tilde{\Pi}_{t+1}B & 0 \\ I & 0 & -\theta N \end{bmatrix} \quad (11.25)$$

This can also be written

$$\Pi_t = G_0 + G_{-1}\begin{bmatrix} 0 & 0 & 0 \\ 0 & 0 & 0 \\ 0 & 0 & \tilde{\Pi}_{t+1} \end{bmatrix} G_1 \quad (11.26)$$

11.2 THE RELATION BETWEEN THE GENERAL AND SPECIAL PROBLEMS

Relation (11.25) and the 'path-Riccati' recursion (11.7) for Π_t indeed imply the Riccati recursions etc. of the control problem. One way of seeing this is noted in Exercise 11.2.1. For another, let us write (11.25) as

$$\Pi_t = \begin{bmatrix} \bar{R}_t & \bar{S}'_t & I \\ \bar{S}_t & \bar{Q}_t & 0 \\ I & 0 & -\theta N \end{bmatrix} \quad (11.27)$$

implying definitions of $\bar{R}_t$, etc. Now, it follows from (11.22) that

$$\min_u F_t(\zeta) = \min_u (\zeta' \Pi_t \zeta) = x' \Pi_t x + 2\lambda' x - \theta \lambda' N \lambda \quad (11.28)$$

and that the minimizing value is $u = K_t x$. Inserting expression (11.27) for Π_t into (11.28) we thus deduce the identifications

$$\begin{aligned} K_t &= -(\bar{Q}^{-1}\bar{S})_t \\ \Pi_t &= (\bar{R} - \bar{S}'\bar{Q}^{-1}\bar{S})_t \end{aligned} \quad (11.29)$$

which are indeed just reformulations of the known relations (c.f. Theorem 7.3.2). It is useful to interpret the recursion

$$\Pi_t \zeta_t + G_1 \zeta_{t-1} = 0 \quad (11.30)$$

for the optimal path in terms of the variables for the control problem.

Theorem 11.2.2. *Recursion* (11.30) *implies the recursions*

$$\begin{aligned} x_t &= x_t^\circ \\ u_t &= K_t x_t^\circ \\ \lambda_t &= -\Pi_t x_t^\circ \end{aligned} \quad (11.31)$$

for the optimally controlled process. Here

$$x_t^\circ = (I + \theta N \Pi_t)^{-1}(Ax_{t-1} + Bu_{t-1}) \quad (11.32)$$

In the limit of constant Π *stability of* Γ *is equivalent to stability of the predictive gain matrix* $\tilde{\Gamma}$ *of equation* (7.33).

Proof. In view of (11.27), relation (11.30) amounts to

$$\begin{bmatrix} \bar{R} & \bar{S}' & I \\ \bar{S} & \bar{Q} & 0 \\ I & 0 & -\theta N \end{bmatrix} \begin{bmatrix} x \\ u \\ \lambda \end{bmatrix}_t = \begin{bmatrix} 0 \\ 0 \\ Ax_{t-1} + Bu_{t-1} \end{bmatrix} \quad (11.33)$$

Solving system (11.33) and appealing to the identities (11.29), we deduce the relations (11.31). These express $(x, u, \lambda)_t$ wholly in terms of the single vector x_t°, which itself obeys the recursion

$$x_{t+1}^\circ = (I + \theta N\Pi)^{-1}(A + BK)x_t^\circ = \tilde{\Gamma}x_t^\circ$$

whence we deduce the final assertion. □

The quantity x_t° can be regarded as an estimate of x_t based upon information available after u_{t-1} has been determined.

11.2.1 Exercises and comments

(1) Write the relations (11.25) or (11.26) as $\Pi_t = H(\tilde{\Pi}_{t+1})$. Then we see by comparing (11.26) with (11.7) that

$$\tilde{\Pi}_t = -H(\tilde{\Pi}_{t+1})^{(33)} \tag{11.34}$$

where $P^{(33)}$ indicates the (33) term in the appropriately partitioned form of the inverse, P^{-1}. Relation (11.34) is indeed just the Riccati recursion

$$\tilde{\Pi}_t = \tilde{f\!f}\tilde{\Pi}_{t+1} \tag{11.35}$$

(c.f. equation (7.27)).

11.3 FORMS OF CANONICAL FACTORIZATION

Formulae (11.15) and (11.25) already supply one form of the canonical factorization

$$\Phi = \Phi_-\Phi_0\Phi_+ \tag{11.36}$$

for the Markov control problem. This is

$$\bar{\Phi}_- = \Phi_+ = \Pi + G_1 z = \begin{bmatrix} \bar{R} & \bar{S}' & I \\ \bar{S} & \bar{Q} & 0 \\ I - Az & -Bz & -\theta N \end{bmatrix} \tag{11.37}$$

$$\Phi_0 = \Pi^{-1} = \begin{bmatrix} \bar{R} & \bar{S}' & I \\ \bar{S} & \bar{Q} & 0 \\ I & 0 & -\theta N \end{bmatrix}^{-1} \tag{11.38}$$

where

$$\bar{R} = R + A'\tilde{\Pi}A = R + A'(\Pi^{-1} + \theta N)^{-1}A \tag{11.39}$$

etc.

If we perform the transformation $\Phi_+ \to P\Phi_+$, $\Phi_0 \to (P')^{-1}\Phi_0 P^{-1}$, with

$$P = \begin{bmatrix} I & -\bar{S}\bar{Q}^{-1} & 0 \\ 0 & \bar{Q}^{-1} & 0 \\ 0 & 0 & I \end{bmatrix} = \begin{bmatrix} I & K' & 0 \\ 0 & \bar{Q}^{-1} & 0 \\ 0 & 0 & I \end{bmatrix} \tag{11.40}$$

11.3 FORMS OF CANONICAL FACTORIZATION

we obtain (appealing to relations (11.29)) the alternative factors

$$\bar{\Phi}_- = \Phi_+ = \begin{bmatrix} \Pi & 0 & I \\ -K & I & 0 \\ I - Az & -Bz & -\theta N \end{bmatrix} \quad (11.41)$$

$$\Phi_0 = \begin{bmatrix} \Pi & 0 & I \\ 0 & \bar{Q}^{-1} & 0 \\ I & 0 & -\theta N \end{bmatrix}^{-1} = \begin{bmatrix} \Pi^{-1}\tilde{\Pi}\theta N & 0 & \Pi^{-1}\tilde{\Pi} \\ 0 & \bar{Q} & 0 \\ \tilde{\Pi}\Pi^{-1} & 0 & -\tilde{\Pi} \end{bmatrix} \quad (11.42)$$

When we come to the higher-order case we shall see that the factorization (11.37), (11.38) is, in many ways, the natural one. It also gives an explicit expression for the optimal control rule in its open-loop form. The recursion

$$\Phi_+ \zeta_t = 0 \quad (11.43)$$

with Φ_+ given by (11.37) is just the recursion $\Pi \zeta_t + G_1 \zeta_{t-1} = 0$ derived from the dynamic programming equation. As such it gives the optimal rule in realizable and open-loop form. The second relation of (11.43) is

$$\bar{S} x_t + \bar{Q} u_t = 0$$

which gives the optimal control rule, and is indeed equivalent to $u_t = K x_t$.

Relation (11.43) with Φ_+ given by (11.41) implies the recursions in their familiar forms:

$$\lambda_t = -\Pi x_t$$
$$u_t = K x_t$$
$$x_t = A x_{t-1} + B u_{t-1} + \theta N \lambda_t$$

The final relation is just the plant equation with predicted noise term $\varepsilon_t = \theta N \lambda_t$.

11.3.1 Exercise and comments

(1) In the case of backward optimization one must achieve a factorization $\Psi_+ \Psi_0 \Psi_-$ of

$$\Psi = \begin{bmatrix} N & L & \mathscr{A} \\ L' & M & \mathscr{C} \\ \bar{\mathscr{A}} & \bar{\mathscr{C}} & -\theta R \end{bmatrix}$$

where $\mathscr{A} = I - Az$, $\mathscr{C} = -Cz$. The analogues of the factors (11.41), (11.42) are then

$$\bar{\Psi}_+ = \Psi_- = \begin{bmatrix} V & 0 & I \\ H & I & 0 \\ \bar{\mathscr{A}} & \bar{\mathscr{C}} & -\theta R \end{bmatrix} \quad \Psi_0 = \begin{bmatrix} V & 0 & I \\ 0 & \overline{M}^{-1} & 0 \\ I & 0 & -\theta R \end{bmatrix}^{-1}$$

where $\overline{M} = M + C\tilde{V}C'$.

11.4 CONDITIONS FOR HORIZON STABILITY AND FACTORIZABILITY

The general formulation of Section 11.1 has such striking simplicity that one is tempted to stay with it, developing conditions for convergence of the Riccati recursion (11.7) with increasing time-to-go and for validity of the factorization (11.15) as a canonical factorization. However, in order to do this we certainly have to supply some ideas about the desired nature of the extremum of the path integral. That is, with respect to which components of ζ should it be minimal? Maximal? Why?

If we revert to the control context then all these questions are answered. We know the nature of the desired extremum, and the ultimate motivation is the minimization of the criterion function $\gamma_\pi(\theta)$.

Theorem 11.4.1. Π_t *has an infinite horizon limit* Π *and* $\Phi(z)$ *possesses the canonical factorizations asserted in Theorem 11.1.2 and Section 11.3 under the conditions of Theorem 9.2.1, which ensure that* $\tilde{\Pi}_t$ *has an infinite-horizon limit.*

Proof. We simply appeal to the fact that Π_t can be expressed in terms of $\tilde{\Pi}_{t+1}$ by relation (11.26), and that the Riccati recursions for the two quantities are equivalent. Convergence of $\tilde{\Pi}_t$ thus implies convergence of Π_t, so that the factorization (11.16) (or, equivalently, (11.15)) undoubtedly holds. Since $\tilde{\Gamma}$ is a stability matrix, then so is Γ, by Theorem 11.2.2. The factorization (11.15) is thus canonical, as are those derived from it in Section 11.3. □

We recall one of the principal conditions of Theorem 9.2.1, that

$$J = J_0 + \theta N = BQ^{-1}B' + \theta N \tag{11.44}$$

should be non-negative definite, and should be positive definite on $\{(A')^t\}$. We recall also the risk-resistance condition

$$\Pi^{-1} + \theta N > 0 \tag{11.45}$$

which must be satisfied by the equilibrium Π, at least.

We have a complete characterization of necessary and sufficient conditions for horizon stability in the scalar case (Theorem 9.1.3). It would be interesting to see what these amount to when seen as conditions for factorizability. We assume that $S = 0$, $R > 0$, $Q > 0$, so that all conditions can be seen as conditions on θ. The condition $\theta > \theta_1$ (c.f (9.4)) implies that $J > 0$. The condition $\theta > \theta_2$ (c.f. (9.5)) implies that the Riccati equation for Π has a finite solution. The condition $\theta > \theta_3$ (c.f. (9.8)), if applicable, enforces the risk-resistance condition (11.45).

Theorem 11.4.2. (i) *Suppose* $|A| \geq 1$. *Then violation of the condition* $\theta > \theta_1$ *implies that* Π_t *(and so* $\tilde{\Pi}_t$*) becomes indefinitely large with increasing horizon.*

(ii) *Suppose $|A| < 1$. Then violation of the condition $\theta > \theta_2$ implies that $|\Phi(z)|$ has a zero on the unit circle.*

(iii) *As the risk-resistance condition (11.45) fails, so the constant factor Φ_0 in the factorization given by (11.38) or (11.42) becomes infinite.*

All these amount to a failure of factorization in that some component is no longer finite, or, as in (ii), that $|\Phi_+|$ and $|\Phi_-|$ must have zeroes on the unit circle, and so are not canonical.

Proof. Assertion (i) is evident. Assertion (iii) follows from the fact that condition (11.45) is equivalent to requiring that θ should exceed the largest value θ_3 at which

$$\begin{bmatrix} \Pi & I \\ I & -\theta N \end{bmatrix}$$

is singular. However, this matrix and the expression for Φ_0^{-1} given by (11.39) are singular or non-singular together. The same is true for the alternative Φ_0^{-1} given by (11.38), which differs from that given by (11.42) only by non-singular factors.

To establish assertion (ii) is a matter of direct calculation. Let us parametrize in terms of J rather than θ. It can be verified that

$$|\Phi(z)| = QR[J + R^{-1}(I - Az)(I - Az^{-1})] \tag{11.46}$$

so that the quadratic equation $|\Phi(z)| = 0$ becomes

$$z + z^{-1} = \frac{1 + A^2 + JR}{A} \tag{11.47}$$

Equation (11.47) will have coincident or complex roots (necessarily then on the unit circle) if the quantity on the right is not greater than 2 in magnitude. In the case $A > 0$ we will thus require that $1 + A^2 + JR > 2A$, or

$$J > -(1 - A)^2/R$$

which is equivalent to $\theta > \theta_2$. The case $A < 0$ follows correspondingly. $\square$

11.5 POLICY IMPROVEMENT

Policy improvement in the general formulation of Section 11.1 would follow just the pattern of the control version in Sections 3.8 and 9.4. We may assume ourselves in the homogeneous case $\omega = 0$, since this will not affect the evaluation of Π.

We seek then to solve

$$\Pi = f\Pi \tag{11.48}$$

where

$$f\Pi = \underset{\Gamma}{\text{stat}}\, [G_0 + G_{-1}\Gamma + \Gamma'G_1 + \Gamma'\Pi\Gamma] = G_0 - G_{-1}\Pi^{-1}G_1 \quad (11.49)$$

Note that the stationary value in (11.49) is achieved by

$$\Gamma = -\Pi^{-1}G_1 \quad (11.50)$$

Specification of Γ amounts to choice of a stationary policy in which ζ_t follows the recursion

$$\zeta_t = \Gamma \zeta_{t-1} \quad (11.51)$$

Suppose Γ determined by (11.50) for a Π which is not necessarily optimal. The linear f-operator corresponding to this fixed policy is f_Π (c.f. Sections 3.8 and 9.4), having the action

$$f_\Pi P = G_0 - 2G_{-1}\Pi^{-1}G_1 + G_{-1}\Pi^{-1}P\Pi^{-1}G_1 \quad (11.52)$$

Policy improvement would then generate a sequence of matrices Π_i and corresponding policies with $\Gamma_{i+1} = -\Pi_i^{-1}G_1$ by solving

$$\Pi_{i+1} = f_{\Pi_i}\Pi_{i+1} \quad (11.53)$$

for Π_{i+1}.

As in Section 3.8, we can assert that this iterative procedure is equivalent to solution of the equilibrium optimality equation (11.48) by the Newton–Raphson method. However, we cannot yet make the assertion usual for policy improvement, that the sequence $\{\Pi_i\}$ will decrease monotonely to the solution of (11.48). This is because the variational nature of the stationary point in (11.49) is unclear, and so the whole problem does not have the usual variational underpinning. If the path integral is that derived from the control problem then we indeed have such an underpinning. It will be found that policy improvement applied in the general path-integral formulation does *not* correspond directly to direct application of policy improvement to the control problem. However, the control motivation nevertheless enables us to establish monotone convergence.

If we are to follow the pattern of Section 9.4 then we require to establish that (a) there exists a class $\mathcal{R}$ of matrices Π satisfying

$$f\Pi \leq \Pi \quad (11.54)$$

and some analogue of the risk-resistance condition, and (b) that for Π and P in $\mathcal{R}$ one can assert that f_Π is monotone, and that

$$fP \leq f_\Pi P \quad (11.55)$$

With these properties we can establish, as in Section 9.4, that, if Π_i lies in $\mathcal{R}$, then $f_{\Pi_i}^s \Pi_i$ also lies in $\mathcal{R}$, and decreases with increasing s to a limit Π_{i+1}, which also lies in $\mathcal{R}$. This establishes the validity of the policy-improvement algorithm in the general setting.

11.5 POLICY IMPROVEMENT

Some hypotheses are required, and are supplied if we indeed assume that we are dealing with the path-integral formulation of the control problem, when we know that we have the $\Pi/\tilde{\Pi}$ relationship

$$\Pi = G_0 + G_{-1} \begin{bmatrix} 1 & 0 & 0 \\ 0 & 1 & 0 \\ 0 & 0 & \tilde{\Pi} \end{bmatrix} G_1 \qquad (11.56)$$

(c.f. (11.26)). Let us write this as

$$\Pi = L(\tilde{\Pi}) \qquad (11.57)$$

Lemma 11.5.1. *If* Π, $\tilde{\Pi}$ *are corresponding in that* (11.57) *holds, then*

$$f\Pi = L(\tilde{f}\tilde{\Pi}) = L(\chi\tilde{\Pi}) \qquad (11.58)$$

Proof. This follows from the fact that $f\Pi$ and $\chi\tilde{\Pi}$ are then also corresponding. □

Lemma 11.5.2. *Define* $\mathscr{R}$ *as* $L(\tilde{\mathscr{R}})$, *so that* $\mathscr{R}$ *is non-empty if* $\tilde{\mathscr{R}}$ *is. Then, for* Π *in* $\mathscr{R}$, Π *is monotonic increasing in* $\tilde{\Pi}$, *also*

$$f\Pi \leq \Pi \qquad (11.59)$$

and $f\Pi$ *lies in* $\mathscr{R}$.

Proof. Recall that $\tilde{\mathscr{R}}$ is the set of $\tilde{\Pi} \geq 0$ for which $\chi\tilde{\Pi} \leq \tilde{\Pi}$ and

$$\tilde{\Pi} = (\Pi^{-1} + \theta M)^{-1} \geq 0 \qquad (11.60)$$

We see from (11.56) that Π is monotonic increasing in $\tilde{\Pi}$, and, under condition (11.60), that $\tilde{\Pi}$ is monotonic increasing in Π, The first assertion then follows. Thus if $\Pi \in \mathscr{R}$ we have $\chi\tilde{\Pi} \in \tilde{\mathscr{R}}$ and

$$f\Pi = L(\chi\tilde{\Pi}) \leq L(\tilde{\Pi}) = \Pi$$

whence the remaining assertions follow. The fact that the risk-resistance condition (11.60) holds for Π in $\mathscr{R}$ is reflected in the general formulation by the fact that the Π of $\mathscr{R}$ are always non-singular (c.f. Theorem 11.42 (iii)). □

One might now imagine that we could prove the validity of policy improvement in the general formulation just by exploiting relation (11.56) and the validity of policy improvement in the control formulation. However, we cannot follow this course, because the policy-improvement algorithms deduced by the two processes are not in complete correspondence. The difficult point is found to be the establishment of (11.55), for which we require the one substantial lemma.

158 THE MARKOV CASE: RECURSIONS AND FACTORIZATIONS

Lemma 11.5.3. *Inequality* (11.55) *holds for all* Π *and* P *of* $\mathscr{R}$ *if*

$$[A \quad B]\begin{bmatrix} R + (\theta N)^{-1} & S' \\ S & Q \end{bmatrix}^{-1}\begin{bmatrix} A' \\ B' \end{bmatrix} > 0 \qquad (11.61)$$

In the case when S has been normalized to zero condition (11.61) reduces to

$$BQ^{-1}B' + \theta AN(I + \theta RN)^{-1}A' > 0 \qquad (11.62)$$

This is not quite the same as the familiar condition

$$J = BQ^{-1}B' + \theta N > 0 \qquad (11.63)$$

of Lemma 9.4.1, although the two do agree in the continuous-time limit.

We shall find it simpler to adopt the system notation of Section 10.7, in which we would write relations (11.25), or (11.56), as

$$\Pi = \begin{bmatrix} \mathscr{R} + \mathfrak{A}'\tilde{\Pi}\mathfrak{A} & \mathfrak{A}'_0 \\ \mathfrak{A}_0 & -\theta N \end{bmatrix} \qquad (11.64)$$

Proof. Since Π and P both belong to $\mathscr{R}$ we can write $\Pi = L(\Pi)$, $P = L(p)$, say. We have

$$\mathbf{f}_\Pi P - \mathbf{f} P = G_{-1}[\Pi^{-1}P\Pi^{-1} - 2\Pi^{-1} + P^{-1}]G_1$$

so that (11.55) holds if

$$\Pi^{-1}P\Pi^{-1} - 2\Pi^{-1} + P^{-1} > 0$$

or if

$$(\Pi - P)P^{-1}(\Pi - P) > 0 \qquad (11.65)$$

However, since Π and P are both of the form (11.64) then (11.65) will hold if

$$[\mathfrak{A} \quad 0]P^{-1}\begin{bmatrix} \mathfrak{A}' \\ 0 \end{bmatrix} > 0 \qquad (11.66)$$

Let us write the left-hand member of (11.66) as z, and define x, y as the solution of the equation

$$P\begin{bmatrix} x \\ y \end{bmatrix} = \begin{bmatrix} \mathfrak{A}' \\ 0 \end{bmatrix}$$

Then, written out in full, we have the system of equations

$$(\mathscr{R} + \mathfrak{A}'\tilde{p}\mathfrak{A})x + \mathfrak{A}'_0 y = \mathfrak{A}'$$

$$\mathfrak{A}_0 x - \theta N y = 0 \qquad (11.67)$$

$$z = \mathfrak{A}x$$

11.5 POLICY IMPROVEMENT

Suppose, for the moment, that N is non-singular. Substituting for y from the second of these equations into the first we deduce that

$$(\mathfrak{R} + \mathfrak{A}'_0(\theta N)^{-1}\mathfrak{A}_0)x + \mathfrak{A}'\tilde{p}z = \mathfrak{A}'$$

Substituting for x from this equation into the last equation of (11.67) we deduce that

$$z = M(I - \tilde{p}z) \tag{11.68}$$

where

$$\mathbf{M} = \mathfrak{A}(\mathfrak{R} + \mathfrak{A}'_0(\theta N)^{-1}\mathfrak{A}_0)^{-1}\mathfrak{A}'$$

We deduce from (11.68) that

$$z = (M^{-1} + \tilde{p})^{-1}$$

Thus $z > 0$ for all $\tilde{p} \geq 0$ if $M > 0$, which is exactly condition (11.61). This condition remains meaningful even if N is singular; see (11.62). □

Lemma 11.5.4. $f_\Pi P$ *is monotone increasing in P, and is of the form $L(p)$.*

Proof. The first assertion is obvious from (11.52), and the second also, if we substitute the explicit forms (11.20) and (11.21) of the G_j.

Lemma 11.5.5. *Suppose $\Pi \in \mathcal{R}$ and that condition (11.61) holds. Then $f_\Pi^s\Pi$ is non-increasing in s and belongs to $\mathcal{R}$.*

Proof. The first assertion follows from $f_\Pi\Pi = f\Pi \leq \Pi$ and the monotonicity of f_Π. It follows then that $f_\Pi^s\Pi = L(\Pi_s)$, say, where Π_s is descreasing in s, so that the risk-resistance condition continues to hold. It follows also, by appeal to (11.55), that

$$f(f_\Pi^s\Pi) \leq f_\Pi^{s+1}\Pi \leq f_\Pi^s\Pi$$

so that $f_\Pi^s\Pi$ indeed belongs to $\mathcal{R}$. □

We can now pull the threads together.

Theorem 11.5.6. *Assume that policy improvement applied to the path-integral formulation of the control problem yields a sequence of matrices $\{\Pi_i\}$, that $\mathcal{R}$ is non-empty and that condition (11.61) holds. Then, if Π_1 belongs to $\mathcal{R}$, then so do all Π_i, and $\{\Pi_i\}$ decreases to a member Π of $\mathcal{R}$ satisfying (11.48).*

The proof is exactly that of Theorem 9.4.2, and the limit Π must be the optimal Π if (11.48) has a unique solution.

Finally, as in Theorem 9.4.3, we can generate members of $\mathcal{R}$.

Theorem 11.5.7. *Suppose that the policy determined by the application of* **f** *to* Π *is stabilizing, in that the linear equation in* Π^*

$$f_\Pi \Pi^* = \Pi^* \tag{11.69}$$

has a finite solution. Assume the condition (11.61) *satisfied. Then*

$$f\Pi^* \leq \Pi^* \tag{11.70}$$

and Π^* *is of the form* $L(\Pi^*)$.

Proof. Relation (11.70) follows from

$$f\Pi^* \leq f_\Pi \Pi^* = \Pi^*$$

and we see from the form (11.52) of f_Π with the G_j given the evaluations (11.20) and (11.21) for the control case that the solution of (11.69) is of the form $L(\Pi^*)$. □

That is, one stage of policy improvement is successful in that if it produces a stabilizing policy, even if not applied to a member of $\mathscr{R}$, it will produce a matrix Π^* which has all the properties of a member of $\mathscr{R}$ except perhaps risk resistance.

CHAPTER 12

Higher-order models: the general path-integral formalism

We now move to serious consideration of the non-Markov case, in which higher-order lags are involved. In doing so, we find that the economy and simplicity of the basic ideas emerge with striking force if we take the analysis first in the general path-integral formulation of Section 10.5.

12.1 FUTURE OPTIMIZATION WITH pth-ORDER DYNAMICS

We return to the general formulation of Section 10.5, assuming that we have to extremize a path integral

$$\mathbb{I}(\zeta) = \sum_\tau \left[\left(\sum_j \zeta'_\tau G_j \zeta_{\tau-j} \right) - \omega'_\tau \zeta_\tau - \zeta'_\tau \omega_\tau \right] + \text{end effects} \quad (12.1)$$

If t is the current instant then we know from Section 10.3 that optimization may have a different character in the ranges $\tau < t$ and $\tau > t$ (with the instant $\tau = t$ possibly hybrid in character) in that one optimizes with respect to different variables in the two ranges. We shall suppose that, as in Section 10.5, expression (12.1) defines a *forward* optimization problem, in which ζ_τ is known for $\tau \leq t$ and is to be chosen to extremize $\mathbb{I}(\zeta)$ freely in the range $\tau > t$. The formalism in the backward case differs only by a time reversal. We shall revert to the explicit control problem in Chapter 13 and consider the coupling of past and future optimizations in Chapter 14.

We shall also suppose the horizon infinite, so that optimization amounts to solution of the semi-infinite system of stationarity conditions

$$\Phi(\mathcal{T})\zeta^{(t)}_\tau = \omega_\tau \quad (\tau > t) \quad (12.2)$$

where

$$\Phi(z) = \sum_j G_j z^j \quad (12.3)$$

(c.f. Section 10.5). Conditions will be required if this infinite-horizon treatment is to be meaningful. For the moment we shall assume these satisfied; sufficient conditions will be established in the control context.

We shall also suppose pth-order dynamics, so that

$$G_j = 0 \qquad (|j| > p) \tag{12.4}$$

with, as ever,

$$G_{-j} = G'_j \tag{12.5}$$

Rather than writing the canonical factorization of Φ as

$$\Phi(z) = \Phi_-(z)\Phi_0\Phi_+(z) \tag{12.6}$$

we shall, for notational convenience, write it rather as

$$\Phi(z) = \overline{\phi(z)}H\phi(z) \tag{12.7}$$

where, again as ever,

$$\overline{\phi(z)} = \phi(z^{-1})' \tag{12.8}$$

The factorization (12.7) is by no means unique, but natural normalizations will suggest themselves.

We expect that $\phi(z)$ will also be of degree p:

$$\phi(z) = \sum_{j=0}^{p} \phi_j z^j \tag{12.9}$$

One can regard the important characterization of ϕ as the statement that, in the case of identically zero ω, the optimal infinite-horizon path satisfies not only

$$\Phi(\mathcal{T})\zeta_\tau = 0 \qquad (\tau > t) \tag{12.10}$$

but also

$$\phi(\mathcal{T})\zeta_\tau = 0 \qquad (\tau > t) \tag{12.11}$$

and that (12.11) constitutes a stable forward recursion. That is, that if we use (12.11) to calculate ζ_τ forwards in time from arbitrarily specified initial values, then ζ_τ tends to zero with increasing τ (necessarily at an exponential rate, if p is finite).

12.2 VALUE FUNCTIONS AND RECURSIONS

The extremization of a path integral implies as many conditions as there are variables. The two ways to reduce these are by factorization and recursive methods. The first reduces the stationarity conditions (12.2) by canonical factorization of $\Phi(\mathcal{T})$. The second defines a value function (the extreme value of the path integral over $\tau \geq t$ conditional on the values of ζ_τ for $\tau \leq t$) and then establishes a recursion in t for this function. Despite the fact that we have

12.2 VALUE FUNCTIONS AND RECURSIONS

advocated the first approach as superior, to develop the relation between the two approaches is very revealing for both.

Define the instantaneous cost

$$c_t = \zeta'_{t-p+1} G_0 \zeta_{t-p+1} + 2 \sum_{j=0}^{p-1} \zeta'_{t-j} G_{p-j} \zeta_{t-p} - 2\omega'_t \zeta_t \qquad (12.12)$$

whose sum over t yields the path integral (12.1) with a convenient regrouping of summands. If we define the value function

$$F(\zeta_t, \zeta_{t-1}, \ldots, \zeta_{t-p+1}; t) = \underset{\zeta_\tau : \tau > t}{\text{stat}} \left[\sum_{\tau \geq t} c_\tau \right] \qquad (12.13)$$

then this latter will satisfy the optimality equation

$$F(\zeta_t, \zeta_{t-1}, \ldots, \zeta_{t-p+1}; t) = \underset{\zeta_{t+1}}{\text{stat}} \left[c_t + F(\zeta_{t+1}, \zeta_t, \ldots, \zeta_{t-p+2}; t+1) \right] \qquad (12.14)$$

Let us specialize for the moment to the homogeneous case $\omega = 0$; the relation of Φ and its canonical factors to the second-degree terms of F is not affected by the values of the lower-degree terms. If we postulate a finite horizon at which F (which is then the terminal cost) is homogeneous quadratic in its arguments, then it follows by induction from (12.14) that this will be so at all times, and we can write

$$F(\zeta_t, \zeta_{t-1}, \ldots, \zeta_{t-p+1}; t) = \sum_{j=0}^{p-1} \sum_{k=0}^{p-1} \zeta'_{t-j} \Pi_{jk}(t) \zeta_{t-k} \qquad (12.15)$$

say. The optimal 'policy' will be of the p-lag linear form in that the extremal criterion in (12.14) will determine ζ_{t+1} linearly in terms of $\zeta_t, \ldots, \zeta_{t-p+1}$: say

$$\sum_{k=0}^{p} \alpha_k(t+1) \zeta_{t-k+1} = 0 \qquad (12.16)$$

When we substitute the quadratic form (12.15) into (12.14) we obtain a system of Riccati equations for the matrix coefficients $\Pi_{jk}(t)$. These are really very much better written in generating function form. Also, for later purposes, it is very much better if we can separate the effects of assuming an arbitrary linear recursion ('policy') for ζ of the form (12.16), and of choosing this recursion to have the optimal form, as specified by the stationarity condition in (12.14). Suppose, then, that we assume relation (12.16) to hold, but with the matrix coefficients $\alpha_k(t+1)$ specified arbitrarily at the moment, except that $\alpha_0(t+1)$ should be non-singular. We shall continue to write the value function as F, with the quadratic representation (12.15), even although the policy need not be optimal. By the substitution

$$\zeta_{t+1} = -\sum_{k=1}^{p} \alpha_0(t+1)^{-1} \alpha_k(t+1) \zeta_{t-k+1} = \sum_{k=1}^{p} \beta_k(t+1) \zeta_{t-k+1} \quad \text{(say)} \quad (12.17)$$

in the square bracket of (12.14) and equation of coefficients we deduce the relations

$$\Pi_{jk}(t) = [\Pi_{j+1,k+1} + \beta'_j \Pi_{0,k+1} + \Pi_{j+1,0}\beta_k + \beta'_j \Pi_{00}\beta_k]_{t+1}$$
$$(j, k = 0, 1, \ldots, p-2)$$

$$\Pi_{j,p-1}(t) = [G_{p-1-j} + \beta'_j G_p + \Pi_{j+1,0}\beta_{p-1} + \beta'_j \Pi_{00}\beta_{p-1}]_{t+1} \quad (12.18)$$
$$(j = 0, 1, \ldots, p-2)$$

$$\Pi_{p-1,p-1}(t) = [G_0 + \beta'_{p-1} G_p + G'_p \beta_{p-1} + \beta'_{p-1} \Pi_{00}\beta_{p-1}]_{t+1}$$

where the $t+1$ subscript in the right-hand members indicates that all time-dependent quantities are evaluated at time $t+1$.

If we determine the coefficients $\alpha_k(t+1)$ optimally we find that relation (12.16) will have the form

$$\sum_{k=0}^{p-1} \Pi_{0k}(t+1)\zeta_{t-k+1} + G_p \zeta_{t-p+1} = 0 \quad (12.19)$$

Relations (12.18) are certainly not transparent as they stand. Let us define the generating functions

$$F(w, z; t) = \sum_j \sum_k \Pi_{jk}(t) w^j z^k \quad (12.20)$$

$$\Phi(w, z) = G_0 + \sum_{j=1}^{p} G_j z^j + \sum_{j=1}^{p} G_{-j} w^j \quad (12.21)$$

$$\phi(z; t+1) = F(0, z; t+1) + G_p z^p \quad (12.22)$$

$$\alpha(z; t+1) = \sum_{k=0}^{p} \alpha_k(t+1) z^k \quad (12.23)$$

We shall refer to $F(w, z; t)$ as the *value function transform*.

Theorem 12.2.1. (i) *When generating functions are formed, recursions* (12.18) *for the value function F under policy* (12.16) *reduce to the equation*

$$F(w, z; t) = (wz)^{-1}[(wz)^p \Phi(z^{-1}, w^{-1}) + F(w, z) - \phi(w)' \alpha_0^{-1} \alpha(z)$$
$$- \alpha(w)' \alpha_0^{-1} \phi(z) + \alpha(w)' \alpha_0^{-1} \phi_0 \alpha_0^{-1} \alpha(z)]_{t+1} \quad (12.24)$$

for the value function transform. The final subscript indicates that F, α and ϕ all carry the additional argument $t+1$.
(ii) *The optimal choice of α is*

$$\alpha(z; t+1) = \phi(z; t+1) = F(0, z; t+1) + G_p z^p \quad (12.25)$$

With this choice, (12.24) *reduces to*

$$F(w, z; t) = (wz)^{-1}[(wz)^p \Phi(z^{-1}, w^{-1}) + F(w, z) - \phi(w)' \phi_0^{-1} \phi(z)]_{t+1} \quad (12.26)$$

Proof is by patient but routine verification.

12.3 A CENTRAL RESULT: THE RICCATI/FACTORIZATION RELATIONSHIP

Relations (12.18) do reduce somewhat when we insert the optimal evaluation (12.19) of α. However, the relations we are left with, just the Riccati equations for this higher-order case, are still not attractive. Relation (12.26), coupled with the definition (12.22) of ϕ in terms of F, constitutes a very striking reduction. We see in the next section that it has implications which are even more striking.

Note the need to introduce the double generating function (12.21), related to $\Phi(z)$ by

$$\bar{\Phi}(z) = \Phi(z^{-1}, z) \qquad (12.27)$$

12.3 A CENTRAL RESULT: THE RICCATI/FACTORIZATION RELATIONSHIP

If we assume the existence of infinite-horizon limits, in which the *optimal* value function transform $F(w, z; t)$ becomes independent of t, then the assertions of Theorem 12.2.1 become even simpler.

Theorem 12.3.1. (i) *In the infinite-horizon limit the optimal value function transform has the evaluation*

$$F(w, z) = (1 - wz)^{-1}[\phi(w)'\phi_0^{-1}\phi(z) - (wz)^p \Phi(z^{-1}, w^{-1})] \qquad (12.28)$$

in terms of

$$\phi(z) = F(0, z) + G_p z^p \qquad (12.29)$$

(ii) *The equation*

$$\phi(\mathcal{T})\zeta_\tau = 0 \qquad (\tau > t) \qquad (12.30)$$

holds along the optimal ζ-path.

The assertions also begin to have a very clear significance. If we set $w = z^{-1}$ in (12.28) we obtain

$$\bar{\Phi}(z) = \overline{\phi(z)}\phi_0^{-1}\phi(z) \qquad (12.31)$$

This is nothing but a canonical factorization of $\bar{\Phi}$, since the forward recursion (12.30) is stable. We thus deduce

Theorem 12.3.2. (i) *If infinite-horizon limits exist then a canonical factorization* (12.7) *exists, in which $\phi(z)$ is a pth-order polynomial.*
(ii) *This factorization can be given the special form* (12.31), *with $\phi(z)$ having identification* (12.29). *It then has the special features:*

$$\phi_p = G_p \qquad (12.32)$$

$$\phi_0 = \Pi_{00} \text{ is symmetric} \qquad (12.33)$$

$$H = \phi_0^{-1} = \phi(0)^{-1} \qquad (12.34)$$

Note that we find the factor ϕ with the particular characteristics (12.32) and (12.33) by a very direct route: we simply write down the stationarity condition with respect to ζ_{t+1} in the optimality equation (12.14). This yields a linear relation (12.16) whose coefficients have properties (12.32) and (12.33); these coefficients become the coefficients of $\phi(z)$ in the infinite-horizon limit. Note also that the double generating function $F(w, z)$ is determined wholly by the single generating function $\phi(z)$. This must be so, because ϕ determines 'policy' which determines costs. Nevertheless, relation (12.28) is striking, see the comments in section 12.6.

12.3.1 Exercises and comments

(1) Note that the assertion of Theorem 12.3.1, that $F(w, z)$ is determined wholly in terms of $F(0, z)$, is equivalent to the assertion that the quadratic function $F(\zeta_t, \zeta_{t-1}, \ldots, \zeta_{t-p+1})$ is determined wholly in terms of its derivative with respect to ζ_t.

(2) Consider the control problem with $S = 0$, $\theta = 0$ and plant equation $\mathscr{A} x_t = Bu_{t-1}$, so that lags above the first occur only in x. The value function will have the form $F(x_t, x_{t-1}, \ldots, x_{t-p+1}) = \sum\sum x'_{t-j}\Pi_{jk}x_{t-k}$ with generating function $F(w, z) = \sum\sum \Pi_{jk} w^j z^k$.

Define $f(z) = F(0, z) = \sum \Pi_{0k} z^k$ and $f_0 = f(0) = \Pi_{00}$. We also interpret $\mathscr{A}$ as the operator $\mathscr{A}(\mathscr{T}) = \sum_j A_j \mathscr{T}^j$ or the generating function $\mathscr{A}(z) = \sum_j A_j z^j$ according to context. Show, by following the calculations of this and the previous section, that

$$F(w, z) = (1 - wz)^{-1}[Rwz - f(w)'f_0^{-1}f(z)$$
$$+ (f(w) - f_0\mathscr{A}(w))'(f_0 - J)(f(z) - f_0\mathscr{A}(z))] \quad (12.35)$$

and hence that

$$R = \overline{(f - f_0\mathscr{A})}J(f - f_0\mathscr{A}) + \mathscr{A}f + \bar{f}\mathscr{A} - \overline{\mathscr{A}}f_0\mathscr{A} \quad (12.36)$$

(all generating functions having argument z). Relation (12.36) has more the character of a Riccati equation for $f(z)$ than of a canonical factorization. This is because, in considering this special case, we have lost the full (x, u, λ) formulation, and so relation (12.36) is in fact the reduced form of a canonical factorization (c.f. Section 13.2).

12.4 POLICY IMPROVEMENT AND THE ITERATIVE DETERMINATION OF A CANONICAL FACTORIZATION

In considering infinite-horizon behaviour we would certainly wish to determine the canonical factor $\phi(z)$, and perhaps also the limiting value function transform $F(w, z)$. We know from (12.28) and (12.29) that either of these quantities is determined by the other.

12.4 POLICY IMPROVEMENT AND THE ITERATIVE DETERMINATION

Iteration of the recursion (12.26) is essentially value iteration. One can expect that the $F(w, z; t)$ thus determined converges to $F(w, z)$ with decreasing t, although conditions ensuring this have yet to be determined; see Chapter 13.

However, to use this recursion seems objectionable on three counts: (1) It is slow—first-order convergence at best; (2) it is inelegant, in that a numerical computation will have to return to the explicit and unenlightening recursions (12.18); (3) it is uneconomic, in that one is determining the double generating function $F(w, z)$ when all that one wants is the single generating function $\phi(z)$.

All these objections are remedied in the following alternative. Let the stage of iteration be labelled by $i = 0, 1, 2, \ldots$, and let

$$\phi_{(i)}(z) = \sum_{j=0}^{p} \phi_{ij} z^j \qquad (12.37)$$

be the determination of $\phi(z)$ at stage i. Consider the procedure in which $\phi_{(i+1)}$ is determined from $\phi_{(i)}$ by the linear equation system

$$\bar{\phi}_{(i+1)} \phi_{i0}^{-1} \phi_{(i)} + \bar{\phi}_{(i)} \phi_{i0}^{-1} \phi_{(i+1)} = \bar{\phi}_{(i)} \phi_{i0}^{-1} \phi_{i+1,0} \phi_{i0}^{-1} \phi_{(i)} + \Phi \qquad (12.38)$$

Theorem 12.4.1. *Recursion* (12.38) *has the following properties*:
(i) *It is the recursion that is generated if the Newton–Raphson algorithm is applied to the factorization*

$$\Phi(z) = \overline{\phi(z)} \phi_0^{-1} \phi(z) \qquad (12.39)$$

regarded as an equation for ϕ.
(ii) *It is also the recursion that would be derived by the application of the policy-improvement algorithm.*
(iii) *It conserves properties* (12.32) *and* (12.33): *that ϕ_{i0} is symmetric, and $\phi_{ip} = G_p$.*

The equivalence of the Newton–Raphson and policy-improvement algorithms is a transference of the equivalence already established in Section 3.8. We may expect, from either of these characterizations, that the method will show second-order convergence (although even convergence in itself has yet to be established; see Section 13.4). However, one striking implication of assertion (ii) above is that the policy-improvement algorithm reduces to a recursion for the canonical factor $\phi(z)$, rather than for the full value function $F(w, z)$. This reinforces the view that policy improvement provides a technique which is natural for the problem.

Proof. To establish assertion (i), write ϕ as $\phi_{(i)} + \Delta_{(i)}$ in (12.39) and retain terms up to first order in Δ. This is exactly the Newton–Raphson procedure. The resulting equation is

$$\Phi = [\bar{\phi} \phi_0^{-1} \phi + \bar{\Delta} \phi_0^{-1} \phi + \bar{\phi} \phi_0^{-1} \Delta - \bar{\phi} \phi_0^{-1} \Delta_0 \phi_0^{-1} \phi]_{(i)} \qquad (12.40)$$

where the (i) subscript indicates that all terms in the bracket should bear this subscript, and Δ_{i0} is the absolute term in the polynomial $\Delta_{(i)}(z)$. Setting $\Delta_{(i)} = \phi_{(i+1)} - \phi_{(i)}$ in (12.40) we deduce equation (12.38) for $\phi_{(i+1)}$, establishing (i).

To establish assertion (iii), suppose that properties (12.32) and (12.33) hold at stage i, so that ϕ_{i0} is symmetric and $\phi_{ip} = G_p$. Equation of coefficients of z^p in (12.38) yields $\phi_{i+1,p} = G_p$, so the second property is certainly conserved. If we set z equal to any value on the unit circle for which $\phi_{(i)}(z)$ is non-singular then equation (12.38) takes the form

$$\alpha = \beta^\dagger \phi_{i+1,0} \beta$$

where α is Hermitian and β non-singular. Thus $\phi_{i+1,0}$ is symmetric, and both properties are indeed retained.

Assertion (ii) is the substantial one: that recursion (12.38) provides the determination of $\phi_{(i+1)}$ in terms of $\phi_{(i)}$ that would follow from policy improvement. We could simply argue that our pth-order model could formally be reduced to a Markov one, and that assertion (2) has been demonstrated for the Markov model in Section 3.8. However, a direct demonstration is satisfying, and provides a good exercise in the use of the generating function formalism.

Choice of 'policy' at stage i is just a choice of the relation

$$\phi_{(i)}(\mathcal{T})\zeta_\tau = 0 \qquad (12.41)$$

determining ζ in terms of earlier values. This must be a stable recursion, so $\phi_{(i)}(z)$ must be chosen to have the canonical property (that its determinant has no zeros in $|z| \leq 1$). Let $F_{(i)}(w, z)$ be the infinite-horizon value function transform under this policy. By relation (12.24) we have

$$F_{(i)}(w, z) = (1 - wz)^{-1}[\psi(w)'\phi_0^{-1}\phi(z) + \phi(w)'\phi_0^{-1}\psi(z) \\ - \phi(w)'\phi_0^{-1}\psi_0\phi_0^{-1}\phi(z) - (wz)^p\Phi(z^{-1}, w^{-1})]_{(i)} \qquad (12.42)$$

where

$$\psi_{(i)}(z) = F_{(i)}(0, z) + G_p z^p \qquad (12.43)$$

However, by Theorem 12.2.1 (ii) we know that one stage of policy improvement would lead us to the revised policy expressed by

$$\phi_{(i+1)}(z) = \psi_{(i)}(z)$$

Setting $\psi_{(i)} = \phi_{(i+1)}$ in (12.42) and setting also $w = z^{-1}$ we deduce relation (12.38). □

The reduction of the policy-improvement step to yield a recursion in terms of $\phi_{(i)}(z)$ rather than $F_{(i)}(w, z)$ is analogous to the reduction of the expression (12.28) for $F(w, z)$ to (12.31), the actual canonical factorization. The reduction is fortunate, since $\phi_{(i)}(z)$ specifies both policy and the current approximation to the canonical factor $\phi(z)$.

12.5 THE PI/NR ALGORITHM

We shall speak of the recursion (12.38) as the PI/NR algorithm, to emphasize that it has both a policy-improvement and a Newton–Raphson justification. Newton–Raphson methods for iterative determinination of a canonical factorization have been much discussed in the literature; see the notes in Section 6. However, they seem to be particularly natural in this context, since they have the variational aspect of equivalence to policy improvement. Moreover, there are the natural normalizations noted in Theorem 12.4.1 (iii). Normalization is often a matter settled arbitrarily, but there may be great virtue in a well-motivated choice.

Equation (12.38) constitutes a set of linear equations for the coefficients $\phi_{i+1,j}$ of $\phi_{(i+1)}(z)$, and the literature essentially concerns itself with developing efficient methods for solving this linear system. For example, the authors of the most recent paper listed, Ježek and Kučera (1985), show that their version of (12.38) can be reduced to the scalar case and then the order of the polynomial progressively reduced by a recursive algorithm.

However, in a sense, equation (12.38) can be solved explicitly. Define

$$\rho_{(i)}(z) = \phi_{i0}^{-1}\phi_{(i)}(z) \qquad (12.44)$$

so that $\rho_{(i)}(\mathcal{T})$ is a stable operator with leading term $\rho_{i0} = I$.

Theorem 12.5.1. *The solution of equation* (12.38) *for $\phi_{(i+1)}$ can be written*

$$\phi_{(i+1)} = [\bar{\rho}_{(i)}^{-1}\Phi\rho_{(i)}^{-1}]_+\rho_{(i)} = \phi_{i0}[\bar{\phi}_{(i)}^{-1}\Phi\phi_{(i)}^{-1}]_+\phi_{(i)} \qquad (12.45)$$

Recall that if $g(z) = \sum_{-\infty}^{\infty} g_j z^j$ is an expansion valid on the unit circle, then $[g(z)]_+ = \sum_0^{\infty} g_j z^j$.

Proof. Let us, for simplicity, set $\phi_{(i+1)} = \phi$ and $\rho_{(i)} = \rho$. Then pre-multiplication of equation (12.38) by $\bar{\rho}^{-1}$ and post-multiplication by ρ^{-1} leads to the relation

$$\phi\rho^{-1} + \bar{\rho}^{-1}\bar{\phi} = \bar{\rho}^{-1}\Phi\rho^{-1} + \phi_{i+1,0} \qquad (12.46)$$

However, $\phi\rho^{-1}$ has an expansion wholly in non-negative powers on $|z| = 1$, since $\rho(\mathcal{T})$ is a stable operator. Selecting the non-negative powers in (12.46) we thus deduce that

$$\phi\rho^{-1} + \phi_{i+1,0} = [\bar{\rho}^{-1}\Phi\rho^{-1}]_+ + \phi_{i+1,0}$$

whence (12.45) follows. □

Relation (12.45) indeed supplies a solution which is explicit to within matrix inversions and application of the truncation operator $[\cdot]_+$, and could itself be said to constitute the PI/NR algorithm.

Suppose that $|\rho(z)|$ has simple zeros only, at values z_r, say (all necessarily outside the unit circle). Then $\bar{\rho}^{-1}\Phi\rho^{-1}$ has the partial fraction expansion

$$\bar{\rho}^{-1}\Phi\rho^{-1} = \sum_r \left[\frac{E_r}{z - z_r} + \frac{E'_r}{(1/z) - z_r} \right]$$

where

$$E_r = \lim_{z \to z_r} (z - z_r)(\bar{\rho}^{-1}\Phi\rho^{-1})$$

and solution (12.45) can be written

$$\phi_{(i+1)}(z) = \sum_r \left[\frac{E_r}{z - z_r} - \frac{E'_r}{z_r} \right] \rho_{(i)}(z)$$

Note also that one need only evaluate the coefficients in $[\bar{\rho}^{-1}\Phi\rho^{-1}]_+$ as far as the term in z^p, because expression (12.45) is a polynomial of degree p.

12.5.1 Exercises and comments

(1) A special case of the solution (12.45) of (12.38) in continuous time is the familiar expression

$$\Pi = \int_0^\infty e^{A't} N e^{At} dt$$

for the solution of the equation $A'\Pi + \Pi A = N$. This is valid if A is a stability matrix, in the continuous-time sense.

(2) Demonstrate that, if Φ contains no powers outside the range $[-p, p]$ and $\phi_{(i)}$ is a polynomial of degree p, then the solution (12.45) is indeed also a polynomial of degree p.

12.6 NOTES ON THE LITERATURE

The author believes the material of this chapter to be new in the generality given—if not, it has certainly not greatly penetrated the literature. The relation between the recursive and factorization approaches has been demonstrated many times, in the control context, but always for the Markov case, and certainly always for the risk-neutral one. It is, of course, a virtue of the general path-integral formulation that much of the detail of the control formulation is bypassed as irrelevant for many of the principle conclusions.

Formula (12.28) is interesting. It recalls, for example, the assertion that, if x_t is a scalar satisfying the stable stochastic difference equation of order p

$$\phi(\mathcal{T})x_t = \varepsilon_t \qquad (12.47)$$

for white noise ε then the elements $v^{(jk)}$ of the inverse of the covariance matrix of

$(x_0, x_1, \ldots, x_{n-1})$ have a generating function

$$\sum_0^{n-1} \sum_0^{n-1} v^{(jk)} w^j z^k = N^{-1} \frac{\phi(w)\phi(z) - (wz)^n \phi(w^{-1})\phi(z^{-1})}{1 - wz}$$

if $n \geq p$ (see Whittle, 1963, p. 73). Positive-definiteness of the quadratic form thus defined provides the classic Hermite condition for stability of system (12.47) (see e.g. Lehnigk, 1966).

There is a very large literature on the computation of the canonical factors of matrix generating functions of the form $\Phi(z) = \sum_{-p}^{p} G_j z^j$. Much of this is algebraic in the reductions proposed, and stated for the case $p = 1$. However, the iterative Newton–Raphson method has been proposed and analysed by a number of authors. The initial suggestions were made by Wilson (1969, 1972) and Hewer (1971); these have been developed by Vostrý (1975, 1976) and Ježek and Kučera (1985). All these authors deal, of course, with some recursion of the type of (12.38), although without the normalization of the factors taken there, and which emerged as natural from the path-integral extremization.

It seems likely that the analytic solution proposed in Section 12.5 could be implemented numerically and so supersede current methods. However, that has yet to be demonstrated.

CHAPTER 13

Canonical factorization in the control context

We must now see how the general results of Chapter 12 apply in the control context. Reversion to the control problem brings additional structure, so that it should be possible to determine the form of the canonical factors more closely, and also to determine conditions for horizon stability and factorizability.

13.1 THE FORM OF THE CANONICAL FACTORS

We revert to the control context, but still considering 'future' optimization, and so essentially the optimization of control under the assumption of a completely observed process variable.

We assume pth-order dynamics, in that A_j and B_j are supposed zero for $j > p$ (c.f. (10.4)). It is natural to generalize the cost structure within the pth-order framework by allowing a cost function

$$\mathbb{C} = \sum_t \sum_j \begin{bmatrix} x \\ u \end{bmatrix}_t' \mathfrak{R}_j \begin{bmatrix} x \\ u \end{bmatrix}_{t-j} + \text{end effects} \tag{13.1}$$

Here we assume that

$$\mathfrak{R}_j = \mathfrak{R}'_{-j} = \begin{bmatrix} R_j & S'_{-j} \\ S_j & Q_j \end{bmatrix} \tag{13.2}$$

and suppose that j lies in the range $[-p, p]$. However, we continue to assume that ε is a white-noise process. The matrix $\Phi(z)$ is then of the more general form (10.50) and we have

$$G_0 = \begin{bmatrix} R_0 & S'_0 & I \\ S_0 & Q_0 & 0 \\ I & 0 & -\theta N \end{bmatrix}$$

$$G_j = G'_{-j} = \begin{bmatrix} R_j & S'_{-j} & 0 \\ S_j & Q_j & 0 \\ A_j & B_j & 0 \end{bmatrix} \quad (1 \leq j \leq p) \tag{13.3}$$

with, of course, $G_j = 0$ for other p-values.

The value function for this pth-order control problem will be of the form $F(x_t, \ldots, x_{t-p+1}; u_{t-1}, \ldots, u_{t-p+1}; t)$ and will satisfy an optimality equation

$$F(x_t, \ldots, x_{t-p+1}; u_{t-1}, \ldots, u_{t-p+1}; t)$$
$$= \underset{u_t, \varepsilon_{t+1}}{\text{stat}} \; [c_t + \varepsilon'_{t+1}(\theta N)^{-1}\varepsilon_{t+1} + F(x_{t+1}, \ldots, x_{t-p+2}; u_t, \ldots, u_{t-p+2}; t+1)]$$

(13.4)

where

$$c_t = \sum_{j=0}^{p} (2 - \delta_j) \begin{bmatrix} x \\ u \end{bmatrix}'_t \mathfrak{R}_j \begin{bmatrix} x \\ u \end{bmatrix}_{t-j} \tag{13.5}$$

and x_{t+1} is to be expressed in terms of earlier variables by the plant equation

$$(\mathscr{A}x + \mathscr{B}u)_{t+1} = \varepsilon_{t+1} \tag{13.6}$$

Now, if we recast equation (13.4) so that it becomes the optimality equation (12.14) of the full (x, u, λ) description then we have a direct route to the normalized canonical factor $\phi(z)$. As noted at the end of Section 12.3, we have only to write down the stationarity condition for $(x, u, \lambda)_{t+1}$ and we have a relation in ζ whose coefficients converge to those of $\phi(z)$ in the infinite-horizon limit.

Define the function

$$F_t^* = F^*(x_t, \ldots, x_{t-p+1}; u_t, \ldots, u_{t-p+1}; t)$$
$$= c_t + \underset{\varepsilon_{t+1}}{\text{stat}} \; [\varepsilon'_{t+1}(\theta N)^{-1}\varepsilon_{t+1} + F(x_{t+1}, \ldots, x_{t-p+2}; u_t, \ldots u_{t-p+2}; t+1)] \quad (13.7)$$

where x_{t+1}, where it occurs, is to be expressed in terms of earlier variables by appeal to the plant equation (13.6).

Lemma 13.1.1. *The stationarity conditions with respect to $\zeta_t = (x, u, \lambda)_t$ of the general formulation amount to*

$$\frac{1}{2} \frac{\partial F_t^*}{\partial x_t} + \lambda_t = 0$$

$$\frac{1}{2} \frac{\partial F_t^*}{\partial u_t} = 0 \tag{13.8}$$

$$(\mathscr{A}x + \mathscr{B}u - \theta N \lambda)_t = 0$$

This holds whatever the control policy before time t and if an arbitrary pth-order linear policy is followed after time t.

By a 'pth-order linear policy' we mean that u_t is taken as an arbitrary linear function of the current state variable $(x_t, \ldots, x_{t-p+1}; u_{t-1}, \ldots, u_{t-p+1})$.

13.1 THE FORM OF THE CANONICAL FACTORS

Proof. The path integral is

$$\mathbb{I}(x, u, \lambda) = \sum_\tau [c + 2\lambda'(\mathscr{A}x + \mathscr{B}u) - \theta\lambda'N\lambda]_\tau$$

We are interested only in the terms for $\tau \geq t$. Extremizing out λ_τ for $\tau > t$ we are left with a sum

$$[c + 2\lambda'(\mathscr{A}x + \mathscr{B}u) - \theta\lambda'N\lambda]_t + \sum_{\tau>t} [c + \varepsilon'(\theta N)^{-1}\varepsilon]_\tau$$

with ε, x and u related by the plant equation in this latter sum. Extremizing out ε_τ and substituting for u_τ from the control rule for $\tau > t$ we are left with the expression

$$F_t^* + [2\lambda'(\mathscr{A}x + \mathscr{B}u) - \theta\lambda'N\lambda]_t$$

to be extremized with respect to $(x, u, \lambda)_t$. The stationarity conditions are indeed those asserted in (13.8), since $A_0 = I$, $B_0 = 0$. □

Suppose now that we indeed optimize over an infinite horizon and that infinite-horizon limits exist, so that F_t^* is not explicitly dependent upon t. Define generating functions $D_{jk}(z)$ by

$$\frac{1}{2} \frac{\partial F_t^*}{\partial x_t} = D_{11}(\mathscr{T})x_t + D_{12}(\mathscr{T})u_t$$

$$\frac{1}{2} \frac{\partial F_t^*}{\partial u_t} = D_{21}(\mathscr{T})x_t + D_{22}(\mathscr{T})u_t \qquad (13.9)$$

Theorem 13.1.2. (i) *Suppose infinite-horizon limits exist. Then, in the control context, the canonical factor $\phi(z)$ of (12.31) has the form*

$$\phi(z) = \begin{bmatrix} D_{11} & D_{12} & I \\ D_{21} & D_{22} & 0 \\ \mathscr{A} & \mathscr{B} & -\theta N \end{bmatrix} \qquad (13.10)$$

where the entries $D_{jk}(z)$ are related to the infinite-horizon value function by (13.7), (13.9).
(ii) *The approximation to the canonical factor $\phi(z)$ yielded by value iteration or policy improvement will also have the form (13.10), with F_{t+1} being replaced in the definition of F_t^* by the value function which is being improved.*

Proof. These assertions simply follow from the fact that the stationarity conditions yield the ζ-recursion

$$\phi(\mathscr{T})\zeta_t = 0 \qquad (13.11)$$

with ϕ defined as in (13.10), and we know from Theorems 12.2.1 and 12.3.1 that this derivation yields the canonical factor. □

The final relation of (13.11) with ϕ given by (13.10) amounts to a statement of the plant equation, with ε estimated by $\theta N \lambda$ (c.f. (10.17)). The second relation is

$$D_{21}(\mathcal{T})x_t + D_{22}(\mathcal{T})u_t = 0 \tag{13.12}$$

This provides an expression of the optimal control u_t. The expression is in closed-loop form, since it was in fact derived from the optimality equation, even if the $D_{jk}(z)$ are to be determined from a canonical factorization. We shall consider the first relation implied by (13.11) in a moment.

In the Markov case the factorization $\Phi = \bar{\phi}\phi_0^{-1}\phi$ took the form (11.37), (11.38), which we then reduced to (11.41), (11.42), to make the determination of Π and K by the canonical factorization explicit. One could do the same in the higher-order case. Let us denote the absolute term in the polynomial D_{jk} by d_{jk}, so that

$$\phi_0 = \begin{bmatrix} d_{21} & d_{12} & I \\ d_{21} & d_{22} & 0 \\ I & 0 & -\theta N \end{bmatrix}$$

which is indeed symmetric. Define now new factors by $\bar{\Phi}_- = \Phi_+ = P\phi$, $\Phi_0 = (P\phi_0 P')^{-1}$, where

$$P = \begin{bmatrix} I & -d_{12} & 0 \\ 0 & I & 0 \\ 0 & 0 & I \end{bmatrix} \begin{bmatrix} I & 0 & 0 \\ 0 & d_{22}^{-1} & 0 \\ 0 & 0 & I \end{bmatrix}$$

The transformed factors are

$$\Phi_+ = \begin{bmatrix} D_{11} - d_{12}d_{22}^{-1}D_{21} & D_{12} - d_{12}d_{22}^{-1}D_{22} & I \\ d_{22}^{-1}D_{21} & d_{22}^{-1}D_{22} & 0 \\ \mathcal{A} & \mathcal{B} & -\theta N \end{bmatrix} \tag{13.13}$$

$$\Phi_0 = \begin{bmatrix} d_{11} - d_{12}d_{22}^{-1}d_{21} & 0 & I \\ 0 & d_{22}^{-1} & 0 \\ I & 0 & -\theta N \end{bmatrix}^{-1} \tag{13.14}$$

In the relation

$$d_{22}^{-1}[D_{21}x_t + D_{22}u_t] = 0$$

(arguments $\mathcal{T}$ understood for the D-terms) u_t has coefficient I, so this constitutes an explicit solution for the optimal control u_t in open-loop form. In the relation

$$(D_{11} - d_{12}d_{22}^{-1}D_{21})x_t + (D_{12} - d_{12}d_{22}^{-1}D_{22})u_t + \lambda_t = 0$$

u_t has coefficient zero, and this constitutes an assertion of the relation

$$\lambda_t = -\frac{1}{2}\frac{\partial F_t}{\partial x_t}$$

13.2. A REDUCTION

It is useful to revert to the system notation of Section 10.7, in terms of which we write

$$\Phi = \begin{bmatrix} \mathfrak{R} & \bar{\mathfrak{A}} \\ \mathfrak{A} & -\theta N \end{bmatrix} \tag{13.15}$$

and so could write expression (13.10) for the canonical factor as

$$\phi = \begin{bmatrix} \mathfrak{D} & \mathfrak{A}'_0 \\ \mathfrak{A} & -\theta N \end{bmatrix} \tag{13.16}$$

It is a direct matter to verify that the expression (13.16) does indeed satisfy

$$\Phi = \bar{\phi}\phi_0^{-1}\phi \tag{13.17}$$

if $\mathscr{D}$ satisfies

$$\mathfrak{R} = [\bar{\mathfrak{D}} \quad \bar{\mathfrak{A}}]\phi_0^{-1}\begin{bmatrix} \mathfrak{D} \\ \mathfrak{A} \end{bmatrix} \tag{13.18}$$

This is a reduced problem, in that $\mathfrak{D}$ is smaller in size than ϕ. In fact, we can see relation (13.18) as the characterization of a reduced factorization problem.

Theorem 13.2.1. *Define*

$$\kappa(z) = \mathfrak{D}(z) + \mathfrak{A}'_0(\theta N)^{-1}\mathfrak{A}(z) \tag{13.19}$$

Then this is the canonical factor for the reduced factorization

$$\mathfrak{R} + \bar{\mathfrak{A}}(\theta N)^{-1}\mathfrak{A} = \bar{\kappa}\kappa_0^{-1}\kappa \tag{13.20}$$

Proof. We have

$$\phi_0 = \begin{bmatrix} \mathfrak{D}_0 & \mathfrak{A}'_0 \\ \mathfrak{A}_0 & -\theta N \end{bmatrix} = \begin{bmatrix} \phi_{11} & \phi_{12} \\ \phi_{21} & \phi_{22} \end{bmatrix} \tag{13.21}$$

say. Let us correspondingly write $\phi_0^{-1} = (\phi^{jk})$. Then relation (13.18) can be written

$$\begin{aligned}\mathfrak{R} &= \bar{\mathfrak{D}}\phi^{11}\mathfrak{D} + \bar{\mathfrak{D}}\phi^{12}\mathfrak{A} + \bar{\mathfrak{A}}\phi^{21}\mathfrak{D} + \bar{\mathfrak{A}}\phi^{22}\mathfrak{A} \\ &= \overline{(\mathfrak{D} + (\phi^{11})^{-1}\phi^{12}\mathfrak{A})}\phi^{11}(\mathfrak{D} + (\phi^{11})^{-1}\phi^{12}\mathfrak{A}) \\ &\quad + \bar{\mathfrak{A}}(\phi^{22} - \phi^{21}(\phi^{11})^{-1}\phi^{12})\mathfrak{A} \end{aligned} \tag{13.22}$$

However, by general matrix identities (see Exercise 2.4.4) we have

$$(\phi^{11})^{-1}\phi^{12} = -\phi_{12}\phi_{22}^{-1} = \mathfrak{A}'_0(\theta N)^{-1}$$

$$\phi^{22} - \phi^{21}(\phi^{11})^{-1}\phi^{12} = \phi_{22}^{-1} = -(\theta N)^{-1}$$

Thus relation (13.22) does indeed reduce to (13.20). The factor κ is canonical, because the recursion

$$\kappa(\mathcal{T})\begin{bmatrix} x \\ u \end{bmatrix}_t = 0 \tag{13.23}$$

is deduced from $\phi(\mathcal{T})\zeta_t = 0$ by simple elimination of λ, and so is stable if this latter is. □

The theorem is valid as it stands if N is non-singular. If N is singular then it must be understood correctly. For example, $\varepsilon' N^{-1}\varepsilon$ is infinite unless ε is orthogonal to the null space of N. Part of the point of including the variable λ is just that singularity of N then causes no difficulty.

Relation (13.20) has indeed an immediate interpretation. The stress in its original form (i.e. before the importation of the λ-variables) had the form

$$\mathbb{S} = \sum_t \mathfrak{x}'_t(\mathfrak{R} + \bar{\mathfrak{A}}(\theta N)^{-1}\mathfrak{A})\mathfrak{x}_t \tag{13.24}$$

$\mathfrak{x}$ being the system variable $\begin{bmatrix} x \\ u \end{bmatrix}$ and generating functions having the argument $\mathcal{T}$. So, factorization (13.20) corresponds to a canonical expression of the stress

$$\mathbb{S} = \sum_t \eta'_t \kappa_0^{-1} \eta_t \tag{13.25}$$

where the transformation

$$\eta_t = \kappa(\mathcal{T})\mathfrak{x}_t \tag{13.26}$$

is invertible into the past.

We can express the operational content of Theorem 13.2.1 as follows.

Theorem 13.2.2. *Let $\kappa(z)$ be the canonical factor determined by (13.20), κ_0 being the absolute term in $\kappa(z)$ and required to be symmetric. Let $\kappa = (\kappa_{jk}; j, k = 1, 2)$ be the natural partitioning of κ. Then the relation determining the optimal control is*

$$\kappa_{21}(\mathcal{T})x + \kappa_{22}(\mathcal{T})u = 0 \tag{13.27}$$

Proof. We know the relation determining the optimal control to be (13.12). However, we can write relation (13.19) more fully as

$$\kappa = \mathfrak{D} + \begin{bmatrix} I \\ 0 \end{bmatrix} (\theta N)^{-1} [\mathscr{A} \quad \mathscr{B}] = \mathfrak{D} + \begin{bmatrix} (\theta N)^{-1}\mathscr{A} & (\theta N)^{-1}\mathscr{B} \\ 0 & 0 \end{bmatrix}$$

Relation (13.12) thus reduces to (13.27). □

The result presents an interesting return to first principles, although with a conclusion by no means evident initially. The theorem states that we can effectively regard $\exp[-\tfrac{1}{2}(\theta\mathbb{C} + \mathbb{D})] = \exp[-\tfrac{1}{2}\mathbb{S}]$ as providing a joint probability density (when normalized) of X_h and U_{h-1}. Relation (13.26) provides the autoregressive representation of this process, normalized in that the coefficient of x_t equals $\operatorname{cov}(\eta_t)$. The optimal control u is then obtained by 'estimating' u. That is, if in equation (13.26) one sets components of η_t equal to zero for which a component of u_t occurs in the right-hand member, one obtains relations (13.27). These are then the 'maximum likelihood' (actually, minimum stress) equations for u_t.

13.3 CONDITIONS FOR HORIZON STABILITY

It must be confessed that this is one of the least satisfactory parts of our analysis. The general path-integral approach is distinguished by its power and economy, and it is reasonable to expect that there are conditions for its validity of a corresponding character: sharp, elegant and expressible directly in generating function form. The sufficient conditions we develop fall far short of this.

If we eliminate u from the system

$$\Phi(\mathcal{T}) = \begin{bmatrix} R & \bar{S} & \bar{\mathcal{A}} \\ S & Q & \bar{\mathcal{B}} \\ \mathcal{A} & \mathcal{B} & -\theta N \end{bmatrix} \begin{bmatrix} x \\ u \\ \lambda \end{bmatrix}_t = 0 \qquad (13.28)$$

by solving for it from the second relation we obtain the reduced system

$$\Phi_1(\mathcal{T}) \begin{bmatrix} x \\ \lambda \end{bmatrix}_t = 0 \qquad (13.29)$$

where

$$\Phi_1 = \begin{bmatrix} R^* & \bar{\mathcal{A}}^* \\ \mathcal{A}^* & -\mathcal{J} \end{bmatrix} \qquad (13.30)$$

Here

$$\begin{aligned} R^* &= R - \bar{S}Q^{-1}S \\ \mathcal{A}^* &= \mathcal{A} - \mathcal{B}Q^{-1}S \end{aligned} \qquad (13.31)$$

are the values to which R and $\mathcal{A}$ reduce under the normalization of S to zero, and $\mathcal{J}$ is the obvious generalization of the control-power matrix:

$$\mathcal{J} = \mathcal{B}Q^{-1}\bar{\mathcal{B}} + \theta N \qquad (13.32)$$

It should be emphasized that expressions (13.31) and (13.32) are now all generating functions, in the operator $\mathcal{T}$ or the transform variable z, as appropriate. Canonical factorization of Φ has now been reduced to the canonical factorization of Φ_1, in which $\mathcal{B}$, Q and N occur only in the combination $\mathcal{J}$.

One might imagine that sufficient controllability and deviation-sensitivity conditions could be expressed in terms of $\mathcal{J}$, R^* and $\mathcal{A}^*$ alone, as they could be for the Markov case (Theorem 9.2.1). This is not, in general, the case.

Theorem 13.3.1. *Suppose past values of control irrelevant, in that $\mathcal{B}(z) = -Bz$, $Q(z) = Q$, so that $\mathcal{J}$ has the constant value $J = BQ^{-1}B' + \theta N$. Then a sufficient controllability condition is that $J \geq 0$ and that J is positive-definite on $\{A^{(t)\prime}\}$, where $\mathcal{A}(z)^{-1}$ has the formal expansion $\sum_0^\infty A^{(j)} z^j$.*

In other cases the controllability criterion cannot necessarily be expressed in terms of $\mathcal{J}$ and $\mathcal{A}$ alone.

Proof. The first assertion follows just from the corresponding assertion for the Markov case (Theorem 9.2.1) with the state variable $(x_t, x_{t-1}, \ldots, x_{t-p+1})$ adopted to achieve a Markov description.

The second follows by counter-example. Consider the scalar problem already discussed in Section 9.3:

$$x_t = ax_{t-1} + b_1 u_{t-1} + b_2 u_{t-2} + \varepsilon_t$$

with the simple cost function $\sum(Rx_t^2 + Qu_t^2)$. We can indeed consider the risk-neutral case, with $\varepsilon = 0$. Consider the two cases

$$b_1 + b_2 z = \begin{cases} b(1 - az) & \text{(i)} \\ b(z - a) & \text{(ii)} \end{cases}$$

for both of which $\mathcal{J}$ has the evaluation

$$\mathcal{J} = b^2 Q^{-1}(1 - az)(1 - az^{-1})$$

In case (i), the system is uncontrollable, as we have seen in Section 9.3. In case (ii) the system is controllable as long as $|a| \neq 1$ (i.e. as long as the two cases are distinct). Controllability can thus not be characterized purely in terms of $\mathcal{J}$. □

The transfer of the conditions of Theorem 9.2.1 to analogous conditions on generating functions would seem to elude us. Such characterizations there must be for factorizability of Φ, but, for the present, one can only state the sufficient conditions obtained by reverting to a state description and appealing to the conditions of Theorem 9.2.1.

13.4 POLICY IMPROVEMENT

We have considered policy improvement in the risk-sensitive Markov case: in the direct control formulation of Section 9.4 and in the path-integral formulation in Section 11.5. In the control formulation it was proved that policy improvement worked if

$$J = BQ^{-1}B' + \theta N > 0 \tag{13.33}$$

13.4 POLICY IMPROVEMENT

and if the class $\mathscr{R}$ of matrices was non-empty. That is, the class of those matrices Π for which

$$\tilde{ff}\Pi \leq \Pi \tag{13.34}$$

and

$$\tilde{\Pi} = (\Pi^{-1} + \theta N)^{-1} \geq 0 \tag{13.35}$$

By policy improvement's 'working' we mean that, starting from an initial matrix Π_0 in $\mathscr{R}$, it generates a sequence of matrices Π_i in $\mathscr{R}$ which decreases to a matrix Π attaining equality in (13.34). The condition that $\mathscr{R}$ should be non-empty can be regarded as replacing the stronger controllability condition of Theorem 9.2.1: that J should be positive definite on $\{(A')^t\}$.

To obtain the equivalent results in the path integral formulation we required again that $\mathscr{R}$ be non-empty, generating a family of matrices $\mathscr{\tilde{R}} = L(\mathscr{R})$ (c.f. Lemma 11.5.2). Condition (13.33) was replaced, however, by the condition

$$[A \ B]\begin{bmatrix} R + (\theta N)^{-1} & S' \\ S & Q \end{bmatrix}^{-1}\begin{bmatrix} A' \\ B' \end{bmatrix} > 0 \tag{13.36}$$

Under these conditions policy improvement, applied to a matrix Π_0 of $\mathscr{R}$, will produce a sequence of matrices Π_i in $\mathscr{R}$ decreasing to a root of

$$\Pi = f\Pi$$

(Theorem 11.5.6). We know indeed from Theorem 11.5.7 that, if the first step of policy improvement is successful in that it produces a stabilizing policy, then, even if Π_0 does not belong to $\mathscr{R}$, the matrix Π_1 will do so in all but possibly the risk-resistance condition.

We wish to generalize these results to the case of pth-order dynamics. This one can always do formally by reverting to a Markov formulation, but this is a recourse one would like to take as seldom as possible.

In the general pth-order formulation one will have an infinite-horizon value function (not necessarily that for the optimal policy)

$$F(\zeta_t, \zeta_{t-1}, \ldots, \zeta_{t-p+1}) = \sum_{j=0}^{p-1} \sum_{k=0}^{p-1} \zeta'_{t-j}\Pi_{jk}\zeta_{t-k} \tag{13.37}$$

and corresponding value function transform

$$F(w, z) = \sum_j \sum_k \Pi_{jk} w^j z^k \tag{13.38}$$

If an inequality $F^{(1)} \leq F^{(2)}$ holds for all values of the ζ-arguments for two value functions of the form (13.37) then we shall write this as

$$F^{(1)}(w, z) \prec F^{(2)}(w, z) \tag{13.39}$$

in terms of the corresponding value function transforms.

Lemma 13.4.1. *The ordering* (13.39) *has the implication*

$$F^{(1)}(z,z) \leq F^{(2)}(z,z) \tag{13.40}$$

for z on the unit circle.

Proof. The ordering (13.39) is equivalent to the inequality

$$\sum_j \sum_k \xi_j^\dagger (F_{jk}^{(2)} - F_{jk}^{(1)}) \xi_k \geq 0$$

for arbitrary complex vectors ξ_j. Taking $\xi_j = z^j \xi$ for z on the unit circle and arbitrary ξ we deduce (13.40). □

We shall write the relation equivalent to $f\Pi \leq \Pi$ as

$$fF(w,z) \prec F(w,z) \tag{13.41}$$

$fF(w,z)$ being the value function transform for the policy under which one optimizes for one unit of time before reverting to the policy with value function transform $F(w,z)$.

Lemma 13.4.2. *Relation* (13.41) *amounts to*

$$(1 - wz)F(w,z) + (wz)^p \Phi(z^{-1}, w^{-1}) \prec \phi(w)' \phi_0^{-1} \phi(z) \tag{13.42}$$

where

$$\phi(z) = F(0,z) + G_p z^p \tag{13.43}$$

Inequality (13.42) *has the implication*

$$\Phi(z) \leq \overline{\phi(z)} \phi_0^{-1} \phi(z) \tag{13.44}$$

for all z on the unit circle.

Proof. Relation (13.42) follows from the fact that $F(w,z)$ undergoes the transformation (12.26) under one stage of value iteration. Inequality (13.44) then follows from Lemma 13.4.1. □

Inequality (13.44) is of interest, indicating as it does that the canonical factorization of $\Phi(z)$ is the equality case of an inequality, the inequality holding for all ϕ within a class which we shall now determine.

Actually, in the general case we have no basis for expecting consistent orderings such as (13.41), because of the mixed character of the extremum of the path integral. It is only by reverting to the control motivation that we obtain such orderings. We first of all need to generalize the idea of the class of matrices $\mathcal{R}$.

Definition. *The generating function $\phi_{(0)}(z)$ of a recursion*

$$\phi_{(0)}(\mathcal{T})\zeta_t = 0 \tag{13.45}$$

13.4 POLICY IMPROVEMENT

is said to be in class $\mathscr{S}$ if it is of the form (13.10) *with $D(z)$ derived from an infinite-horizon value function for the control problem by relations* (13.7) *and* (13.9).

This just reflects the fact that the relations determining $(x, u, \lambda)_t$ are obtained by one stage of value iteration in the actual control context.

Theorem 13.4.3. *Suppose that the pth-order analogue of condition* (13.36) *holds. Assume that $\phi_{(0)}(z)$ belongs to $\mathscr{S}$ and is stable, in that that* (13.45) *is a stable forward recursion for ζ_t. Consider the sequence of generating functions $\phi_{(i)}$ generated by iterating the policy-improvement procedure* (12.38) *(or* (12.45)*), starting from $\phi_{(0)}$. Then the $\phi_{(i)}$ all belong to $\mathscr{S}$ and are stable, and $\bar{\phi}_{(i)} \phi_{i0}^{-1} \phi_{(i)}$ decreases monotonically (at least for $i > 1$) to $\Phi(z)$ with increasing i.*

Proof. This just a transference of the assertions of Theorems 12.5.6 and 12.5.7 to the pth order case, with the simplification that characterizations in terms of value functions or their transforms can be reduced to characterizations in terms of the approximations $\phi_{(i)}(z)$ to the canonical factor. This is because ϕ determines F and the policy-improvement procedure is couched wholly in terms of ϕ. □

CHAPTER 14

The recoupling of past and future

We come now to the coupling of past and future optimizations. To do this we must revert to the control context, because it is that which tells us which sets of variables are involved in the two optimizations.

In the risk-neutral case coupling amounted to the simple substitution of estimates $x_{t-j}^{(t)}$ ($0 \leq j < p$) based on past data into the expression for complete-observation control $u_t(x_t, \ldots, x_{t-p+1}; u_{t-1}, \ldots, u_{t-p+1})$ based on the optimization of future costs. Even here, though, there is an additional feature in the non-Markov case: that we have to generate not only the estimate $x_t^{(t)}$ of the current process variable but also the estimates of past values.

In the risk-sensitive case coupling is a two-way business, because the estimates $x_{t-j}^{(t)}$ to be inserted in the control law now depend upon future as well as upon past stress. It is the reduction of this calculation that we shall now address.

14.1 THE MATCHING CONDITIONS FOR PAST AND FUTURE OPTIMIZATIONS

We have the two sets of equations, (10.26) and (10.27), which we write in condensed form as

$$\Phi \zeta_\tau^{(t)} = \omega_\tau \quad (\tau \geq t) \tag{14.1}$$

$$\Psi \chi_\tau^{(t)} = \rho_\tau \quad (\tau \leq t) \tag{14.2}$$

If we assume the canonical factorizations

$$\Phi = \bar{\phi} \Phi_0 \phi$$
$$\Psi = \psi \Psi_0 \bar{\psi} \tag{14.3}$$

without, at the moment, settling on any particular normalization, then equations (14.1) and (14.2) can be partially inverted to

$$\Phi_0 \phi \zeta_\tau^{(t)} = \bar{\phi}^{-1} \omega_\tau = \omega_\tau^* \quad (\tau \geq t) \tag{14.4}$$

$$\Psi_0 \bar{\psi} \xi_\tau^{(t)} = \psi^{-1} \rho_\tau = \rho_\tau^* \quad (\tau \leq t) \tag{14.5}$$

where ω_τ^* and χ_τ^* are known, at least for $|\tau - t| \geq p$. With these reductions one is very near a complete solution of the problem. For example, in the risk-neutral case equations (14.5) for $t - p < \tau \leq t$ determine the estimate $x_t^{(t)}$ of the process variable x_τ over the same time interval, and equation (14.4) for $\tau = t$ then determines the optimal control $u_t^{(t)}$ explicitly in terms of these estimates.

This is, then, one way in which the two equation systems (10.4) and (10.5) are coupled; the solution of (10.4) for $\zeta_\tau^{(t)}$ ($\tau \geq t$) depends upon the values of $x_\tau^{(t)}$ ($t - p < \tau \leq t$). In the risk-sensitive case there is also a back coupling, in that the solution of (14.5) for $\chi_\tau^{(t)}$ ($\tau \leq t$) depends upon the values of $\lambda_\tau^{(t)}$ ($t \leq \tau < t + p$). This back coupling reflects the effect that future costs have upon estimates of past process variables in a risk-sensitive situation.

Formally, what one must do is to write down equations (14.4) for $t \leq \tau < t + p$ and equations (14.5) for $t - p < \tau \leq t$, thus determining ζ and χ over the respective intervals. One thus has a system of $2p$ matrix equations, or $p(4n + m + r)$ scalar equations, which one can regard as determining the process variable estimates $x_\tau^{(t)}$ ($t - p < \tau \leq t$) or simply as determining the optimal control $u_t^{(t)}$ in terms of observables.

This is not a very attractive point at which to abandon the problem to computation; one feels that the equation system permits further reduction.

14.1.1 Exercise and comments

(1) Regard the partial inversion (14.5) as indeed a solution of the stable forward recursion $\psi \rho_\tau^* = \rho_\tau$ (ρ being assumed completely known) followed by solution of the stable backward recursion $\Psi_0 \bar\psi \chi_\tau^{(t)} = \rho_\tau^*$ (for $\tau \leq t$). Show that then formally

$$\sum_t \sum_{j \geq 0} \chi_{t-j}^{(t)} z^t w^j = (1 - wz)^{-1} \psi'(w)^{-1} \Psi_0^{-1} \psi(z)^{-1} \sum_t \rho_t z^t$$

where $\psi(w)$, $\psi(z)$ and $(1 - wz)^{-1}$ are to be expanded in non-negative powers of w and z. This may explain the appearance of some of the double generating functions in Sections 12.2 and 12.3.

14.2 THE MARKOV CASE

We know from Section 8.3 how past and future observations should be recoupled in the Markov case: the minimal stress estimate $x_t^{(t)}$ to be inserted in the control rule is given by

$$x_t^{(t)} = (I + \theta V \Pi)^{-1}(\hat{x}_t + \theta V \sigma_t) \tag{14.6}$$

Here $\hat{x}_t$ is the estimate of x_t based upon past stress alone, yielded by the Kalman filter (8.12). Correspondingly, $-2\sigma_t' x_t$ is the term linear in x_t in future stress and σ_t is yielded by the backward recursion (7.45).

It is of interest to see how relation (14.6) would emerge from the matching conditions of the previous section, because it is a considerable reduction of these. Essentially, solution of a system of $4n + m + r$ scalar equations has been reduced to solution of a system of n scalar equations.

Take the past equations first. We see from the form of canonical factorization noted in Exercise 11.3.1 that system (14.5) amounts to

$$\begin{bmatrix} V & 0 & I \\ H & I & 0 \\ \bar{\mathcal{A}} & \bar{\mathcal{C}} & -\theta R \end{bmatrix} \begin{bmatrix} -\theta\lambda \\ -\theta\mu \\ x \end{bmatrix}^{(t)}_\tau = \cdots \tag{14.7}$$

Here ... indicates known terms. We thus have the equation in x and λ alone at $\tau = t$:

$$x_t^{(t)} - \theta V \lambda_t^{(t)} = \cdots \tag{14.8}$$

It is the term in $\lambda_t^{(t)}$ which supplies the coupling of estimate $x_t^{(t)}$ with future stress; if this were missing, equation (14.8) would have to reduce to $x_t^{(t)} = \hat{x}_t$. Therefore, in fact, (14.8) must take the form

$$x_t^{(t)} - \theta V \lambda_t^{(t)} = \hat{x}_t \tag{14.9}$$

Correspondingly, we deduce the relation

$$\Pi x_t^{(t)} + \lambda_t^{(t)} = \sigma_t \tag{14.10}$$

from the forward system. Eliminating λ from relations (14.9) and (14.10) we deduce (14.6).

14.3 THE HIGHER-ORDER CASE

At the moment it would appear that no further reduction is possible in the higher-order case than to appeal to the pth-order version of solution (14.6). Thus, for example, assume for the moment that higher-order lags occur only in x, so that the effective state variable is $(x_t, x_{t-1}, \ldots, x_{t-p+1})$, and suppose that for a given value of this variable the extremized values of past and future stress are

$$F_t = \sum_j \sum_k x'_{t-j} \Pi_{jk} x_{t-k} - 2 \sum_j x'_{t-j} \sigma^{(t)}_{t-j} + \cdots$$
$$P_t = \sum_j \sum_k (x_{t-j} - \hat{x}^{(t)}_{t-j})' V^{jk} (x_{t-k} - \hat{x}^{(t)}_{t-k}) + \cdots \tag{14.11}$$

Here $+ \cdots$ indicates terms independent of x, the V^{jk} are matrix components of the inverse matrix $(V^{jk}) = (V_{jk})^{-1}$, and j and k run over the ranges $0 \leq j, k < p$ in all cases. Then, by appeal to equation (14.6), the minimal stress estimates of x are determined by the equation system

$$x^{(t)}_{t-j} + \theta \sum_l \sum_k V_{jl} \Pi_{lk} x^{(t)}_{t-k} = \hat{x}^{(t)}_{t-j} + \theta \sum_k V_{jk} \sigma^{(t)}_{t-k} \qquad (0 \leq j < p) \tag{14.12}$$

where summations are over the range $0 \leq l, k < p$. Here $\hat{x}_\tau^{(t)}$ is the value of x_τ which extremizes the *past* stress at time t. It is the value of x_τ that one obtains if one solves the system (14.5) while neglecting back coupling, i.e. by assuming $\lambda_\tau^{(t)} = 0$ for $\tau \geq t$; correspondingly for $\sigma_\tau^{(t)}$ for the forward system.

188   THE RECOUPLING OF PAST AND FUTURE

System (14.12) could only be further reduced if the matrix with jkth matrix element $\sum_l V_{jl}\Pi_{lk}$ were perceived to have some special structure. It is possible that it may do so; we know, for example, from Theorem 12.3.1 that the whole matrix (Π_{jk}) is determined by Π_{0k} ($0 \le k < p$). However, such a reduction has yet to be achieved.

14.3.1 Exercise and comments

(1) Generalize relations (14.12) to the case when higher-order lags in u also occur in the model.

CHAPTER 15

Continuous time: the path-integral formalism

15.1 THE MARKOV REGULATION PROBLEM

For the forward optimization of control with the Markov regulation model the path integral becomes

$$\mathbb{I}(x, u, \lambda) = \int [c(x, u) + 2\lambda'(\dot{x} - Ax - Bu) - \theta\lambda'N\lambda]\, d\tau \qquad (15.1)$$

One can either establish (15.1) directly or take it as a limit version of the sum (10.11), with the usual conventions on correspondences (c.f. Section 3.9). The stationarity conditions are

$$\begin{bmatrix} R & S' & -\mathscr{D} - A' \\ S & Q & -B' \\ \mathscr{D} - A & -B & -\theta N \end{bmatrix} \begin{bmatrix} x \\ u \\ \lambda \end{bmatrix} = 0 \qquad (\tau \geq t) \qquad (15.2)$$

where $\mathscr{D}$ denotes the operation of differentiation with respect to time:

$$\mathscr{D}x = \dot{x} \qquad (15.3)$$

We shall write (15.2) as

$$\Phi(\mathscr{D})\zeta = 0 \qquad (15.4)$$

and one can verify that $\Phi(\mathscr{D})$ is the formal limit as $\Delta \downarrow 0$ of $\Delta^{-1}\Phi(e^{-\Delta\mathscr{D}})$, where this is the Φ of the discrete-time case, given by (11.19), and we again make the usual correspondences.

In discrete time we often replaced the operator $\mathscr{T}$ by a scalar variable z, the point being that $\mathscr{T}z^t = z(z^t)$. We shall now often correspondingly replace the operator $\mathscr{D}$ by a complex scalar z, the point being that $\mathscr{D}(e^{zt}) = z(e^{zt})$. It would be better to have a different notation in the two cases, since the variables are not quite corresponding, and a more usual choice in the continuous-time context would be s or $i\omega$. However, economy and the need to avoid conflict lead us to retain z.

As in the discrete-time case (c.f. Section 10.4), the optimal control problem is solved, at least for the infinite-horizon case with perfect state observation and appropriate regularity conditions if we can factorize $\Phi(z)$ canonically:

$$\Phi = \Phi_- \Phi_0 \Phi_+$$

Canonical behaviour will require, as before, that both $\Phi_+(\mathscr{D})$ and its inverse ($\Phi_-(\mathscr{D})$ and its inverse) should operate properly into the past (future). In the case of rational $\Phi(z)$ this is equivalent to requiring that the poles and zeros of $|\Phi_+(\mathscr{D})|$ ($|\Phi_-(\mathscr{D})|$) should lie strictly in the left (right) half of the complex z-plane (c.f. Exercise 15.1.1). The conjugation operator now has the effect $\overline{\Phi(z)} = \Phi(-z)'$. The Φ-operator defined by (15.2) and (15.3) is thus self-conjugate.

A possible factorization of this Φ is

$$\overline{\Phi}_-(z) = \Phi_+(z) = \begin{bmatrix} \Pi & 0 & I \\ S + B'\Pi & Q & 0 \\ z - A & -B & -\theta N \end{bmatrix} \qquad (15.5)$$

$$\Phi_0 = \begin{bmatrix} \theta N & 0 & I \\ 0 & Q^{-1} & 0 \\ I & 0 & 0 \end{bmatrix} \qquad (15.6)$$

where Π is indeed the solution of the equilibrium Riccati equation:

$$R + A'\Pi + \Pi A - (S' + \Pi B)Q^{-1}(S + B'\Pi) - \theta \Pi N \Pi = 0$$

(c.f. (7.49)). This can be verified either directly or as a limit version of the discrete-time factorization; see Exercise 15.1.2.

One could clearly take the slightly modified factors

$$\overline{\Phi}_- = \Phi_+ = \begin{bmatrix} \Pi & 0 & I \\ -K & I & 0 \\ z - A & -B & -\theta N \end{bmatrix}$$

$$\Phi_0 = \begin{bmatrix} \theta N & 0 & I \\ 0 & Q & 0 \\ I & 0 & 0 \end{bmatrix}$$

where $K = -Q^{-1}(S + B'\Pi)$ is indeed the matrix of the optimal control rule $u = Kx$. These factors are the analogues of (11.41), (11.42).

It is not obvious, but is nevertheless true, that the choice of factors (15.5) and (15.6) is indeed the continuous-time analogue of the natural form (11.37), (11.38). To see this, and many other points, it is now best to revert to the general path-integral formulation.

15.1.1 Exercises and comments

(1) The forward recursion $\phi(\mathscr{T})\zeta = 0$ in discrete time is stable if $|\phi(z)|$ has all its zeros strictly outside the unit circle. The forward recursion $\phi(\mathscr{D})\zeta = 0$ in continuous time is stable if $|\phi(z)|$ has all its zeros strictly in the left half-plane.

(2) Consider the factorization $\bar{\phi}\phi_0^{-1}\phi$ for the discrete-time case given in (11.37), (11.38). Show that the Φ_+ of (15.5) is the formal limit as $\Delta \downarrow 0$ of $P\phi(e^{-\Delta z})$ and the Φ_0 of (15.6) is the formal limit of $\Delta^{-1}(P\phi_0 P')^{-1}$, where

$$P = \begin{bmatrix} I & 0 & 0 \\ 0 & \Delta^{-1} & 0 \\ 0 & 0 & \Delta^{-1} \end{bmatrix}$$

15.2 THE GENERAL PATH-INTEGRAL FORMULATION

The continuous-time version of the material of Chapter 12 runs to some degree in parallel with that material, but is divergent enough that we should give an independent treatment. As for the discrete-time case, we consider a forward optimization in which the n-vector variable ζ can be varied freely.

Assume, then, that we are extremizing a path integral

$$\mathbb{I}(\zeta) = \int \mathbf{c}(\zeta) \, d\tau + \text{end effects} \tag{15.7}$$

where c is a quadratic function of ζ and its derivatives:

$$\mathbf{c}(\zeta) = \sum_j \sum_k \sum_r \sum_s c_{jk}^{(rs)} (\mathscr{D}^r \zeta_j)(\mathscr{D}^s \zeta_k) - \omega' \zeta - \zeta' \omega \tag{15.8}$$

Here ζ_j is the jth component of ζ and ω is a known vector function of time; all variables in (15.8) are evaluated at a common time argument, which is expressed only when it need be.

We shall often find it simpler to use the notation

$$\zeta_j^{[r]} = \mathscr{D}^r \zeta_j \tag{15.9}$$

for derivatives, so that c can be written more compactly

$$\mathbf{c}(\zeta) = \sum_j \sum_k \sum_r \sum_s c_{jk}^{(rs)} \zeta_j^{[r]} \zeta_k^{[s]} - \omega' \zeta - \zeta' \omega \tag{15.10}$$

As a matter of normalization we can demand the symmetry

$$c_{kj}^{(sr)} = c_{jk}^{(rs)} \tag{15.11}$$

However, even with this normalization, representation (15.10) is non-unique, in that partial integration of the path integral yields the effective equality

$$\zeta_j^{[r]} \zeta_k^{[s]} = -\zeta_j^{[r-1]} \zeta_k^{[s+1]} \qquad (r > 0) \tag{15.12}$$

The equality is effective in that the difference of the two sides of (15.12) is integrable. Its time integral thus contributes purely to the end effects, and does not affect the optimization at time points interior to the range of integration.

We can carry the reduction (15.12) so far as to write the path integral (15.7) as

$$\mathbb{I} = \int \sum_j \sum_k \sum_r \sum_s c_{jk}^{(rs)} (-)^r \zeta_j \zeta_k^{[r+s]} \, d\tau - 2 \int \zeta' \omega \, d\tau + \text{end effects} \tag{15.13}$$

If we demand that $\mathbb{I}$ be stationary with respect to variations in ζ_j we thus derive the equation

$$\sum_k \sum_r \sum_s c_{jk}^{(rs)}(-)^r \zeta_k^{[r+s]} = \omega_j \qquad (15.14)$$

We can express conclusion (15.14) as

Theorem 15.2.1. *The condition that $\{\zeta(t)\}$ should render the path integral (15.7) stationary can be written*

$$\Phi(\mathscr{D})\zeta = \omega \qquad (15.15)$$

where the $n \times n$ matrix of operators $\Phi(\mathscr{D})$ has jkth element

$$\Phi_{jk}(\mathscr{D}) = \sum_r \sum_s c_{jk}^{(rs)}(-\mathscr{D})^r \mathscr{D}^s \qquad (15.16)$$

This matrix is Hermitian in that

$$\Phi_{jk}(\mathscr{D}) = \Phi'_{kj}(-\mathscr{D}) \qquad (15.17)$$

Equation (15.15) is just a rewriting of (15.14), and the symmetry relation (15.17) follows from (15.11) and (15.16). As ever, we write relation (15.17) as

$$\Phi = \bar{\Phi} \qquad (15.18)$$

Relation (15.15) is, of course, the continuous-time analogue of (12.2).

As in the previous sections, canonical factorization of $\Phi(z)$ will take the form

$$\Phi(z) = \Phi_-(z)\Phi_0\Phi_+(z) \qquad (15.19)$$

where Φ_0 is independent of z, and we may assume, in view of (15.18), that $\Phi_- = \bar{\Phi}_+$. Since Φ is polynomial in z then so also, presumably, are the factors Φ_- and Φ_+. Canonicality will require that

$$\Phi_+(\mathscr{D})\zeta = 0 \qquad (15.20)$$

be a stable forward recursion for ζ. That is, the equation should constitute a valid recursion in that the matrix constituted by the coefficients of derivatives of highest order in each row of (15.20) should be non-singular, and it should be stable in that the equation

$$|\Phi_+(z)| = 0 \qquad (15.21)$$

in the complex scalar z should have all its zeros strictly in the left half-plane (i.e. with strictly negative real part).

Determination of the canonical factorization again determines the infinite-horizon optimal policy, provided regularity conditions are satisfied. The partial inversion

$$\Phi_+(\mathscr{D})\zeta = \Phi_0^{-1}\Phi_-(\mathscr{D})^{-1}\omega \qquad (15.22)$$

of (15.15) expresses the optimal choice of ζ in open-loop and feedback/feedforward form. That is, relation (15.22) gives the optimal determination of current ζ in terms of past ζ and future ω, whatever past policy may have been.

15.2.1 Exercise and comments

(1) The linear term $\zeta'\omega$ or $\omega'\zeta$ in the path integral (15.15) could well have been replaced by a term involving the differentials of ζ: say, $\sum_r \omega_r' \zeta^{[r]}$. The effect of this would simply be to replace ω in the right-hand member of (15.15) by $\sum_r (-)^r \omega_r^{[r]}$.

15.3 RECURSIONS AND FACTORIZATIONS

Relation (15.12) implies some ambiguity in the expression of the 'instantaneous cost' c. This will usually be resolved by the requirement that c have certain definiteness properties as a quadratic form: e.g. positive definite in some components of ζ and negative definite in others.

Suppose that, in this canonical form, the differentials of ζ_j occur in $c(\zeta)$ up to order r_j exactly ($j = 1, 2, \ldots, n$). Then what one might term a linear minimal-order policy would determine the forward evolution of ζ by a set of linear relations

$$\zeta_j^{[r_j]} = \sum_k \sum_{s < r_k} \kappa_{jk}^{(s)} \zeta_k^{[s]} \qquad (15.23)$$

say, where the coefficients κ may be time dependent. These coefficients determine the policy, which we can regard as the higher-order equivalent of a linear Markov policy. The point is that the optimal policy will lie within this class, and determination of the optimal system (15.23) in the infinite-horizon case is equivalent to the determination of the canonical factor $\Phi_+(\mathscr{D})$.

Let $F(\zeta, t)$ denote the value function under policy (15.23), starting from a known ζ-history at time t. Although we have loosely written this as ζ-dependent, in fact it will depend upon the function $\zeta(\tau)$ only through the differentials $\zeta_j^{[r]}$ ($r < r_j$; $j = 1, 2, \ldots, n$) at the current instant t. Therefore if $r_j = 0$ (i.e. if ζ_j occurs in c only in undifferentiated form) then ζ_j will not occur in it at all. Let us assume homogeneity, so that $\omega = 0$, and assume also that the terminal contributions to the path integral are homogeneous quadratic, if the horizon is finite. It follows then that the value function under policy (15.23) is of the homogeneous quadratic form

$$F(\zeta, t) = \sum_j \sum_k \sum_r \sum_s \Pi_{jk}^{(rs)}(t) \zeta_j^{[r]} \zeta_k^{[s]} \qquad (15.24)$$

and F will obey the backward equation

$$c + \frac{\partial F}{\partial t} + \sum_j \sum_r \zeta_j^{[r+1]} \frac{\partial F}{\partial \zeta_j^{[r]}} = 0 \qquad (15.25)$$

with $\zeta_j^{[r_j]}$ expressed in terms of lower-order derivatives by (15.23). If the optimal policy is adopted at time t (although not necessarily either sooner or later) then (15.25) would be replaced by

$$\text{stat}\left[\mathbf{c} + \frac{\partial F}{\partial t} + \sum_j \sum_r \zeta_j^{[r+1]} \frac{\partial F}{\partial \zeta_j^{[r]}}\right] = 0 \quad (15.26)$$

where 'stat' represents the evaluation of the bracket at a point stationary with respect to the highest-order derivatives $\zeta_j^{[r_j]}$ ($j = 1, 2, \ldots, n$) and n is the dimension of ζ.

The relations implied by (15.25) or (15.26) for the coefficients $\Pi_{jk}^{(rs)}$ are the continuous-time versions of relations (12.18) and (12.19), and, like them, not elegant. It is again advantageous to form transforms, and we define the $n \times n$ matrices of generating functions

$$F(w, z) = \left(\sum_r \sum_s \Pi_{jk}^{(rs)} w^r z^s\right)$$

$$c(w, z) = \left(\sum_r \sum_s c_{jk}^{(rs)} w^r z^s\right)$$

$$\kappa(z) = \left(\sum_r \kappa_{jk}^{(r)} z^r\right)$$

the t-dependence of F and Π being understood. We shall have occasion to distinguish the parts of these generating functions associated with terms which are linear or quadratic in the highest-order derivatives, and so define

$$H = (c_{jk}^{(r_j, r_k)})$$

$$d(z) = \text{diag}(z_j^{r_j})$$

$$c_1(z) = \left(\sum_{s < r_k} c_{jk}^{(r_j, s)} z^s\right)$$

$$F_1(z) = \left(\sum_s \Pi_{jk}^{(r_j - 1, s)} z^s\right)$$

Theorem 15.3.1. *The value function transform under policy* (15.23) *satisfies the backward equation*

$$c(w, z) + \frac{\partial F(w, z)}{\partial t} + (w + z)F(w, z) + \kappa(w)'H\kappa(z) - d(w)Hd(z)$$
$$+ [\kappa(w) - d(w)]'[c_1(z) + F_1(z)] + [c_1(w) + F_1(w)]'[\kappa(z) - d(z)] = 0 \quad (15.27)$$

Proof. If there were no need to take account of the fact that the highest-order derivatives are to be expressed in terms of lower orders by (15.23) then relation

15.3 RECURSIONS AND FACTORIZATIONS

(15.25) would lead to the simpler form

$$\mathbf{c} + \frac{\partial F}{\partial t} + (w + z)F = 0$$

of (15.27). The two last terms in the left-hand member of (15.27) represent the effect of making this substitution in the terms linear in highest-order derivatives in (15.25); the two terms before those represent the effect of making this substitution in the quadratic terms. □

It is a simplificaton of the continuous-time case that these quadratic terms arise solely from instantaneous cost **c** and not at all from future cost **F**. The reader who finds our argument unconvincing can appeal to explicit relationships between the coefficients $\Pi_{jk}^{(rs)}$, as we did in Section 12.2.

Corollary 15.3.2. *Assume policy* (15.23) *stationary and stable, so that a time-independent infinite-horizon limit exists. Then relation* (15.27) *implies that*

$$\Phi(z) = \bar{d}Hd - \bar{\kappa}H\kappa + \overline{(d - \kappa)}(c_1 + F_1) + \overline{(c_1 + F_1)}(d - \kappa) \qquad (15.28)$$

where $\Phi(z)$ *is the matrix* (15.16) *of the stationarity conditions* (15.15), *and* $\kappa = \kappa(z)$, $\bar{\kappa} = \kappa(-z)'$, *etc.*

Proof. This follows simply by setting $w = -z$ in the equilibrium form of (15.27), and recognizing that

$$c(-z, z) = \Phi(z) \qquad (15.29)$$

□

Suppose now that the policy is chosen optimally, either at or from the current instant.

Theorem 15.3.3. (i) *If one chooses the currently optimal values of the highest order derivatives of* ζ *then relations* (15.23) *take the form*

$$\phi(\mathcal{D})\zeta = 0 \qquad (15.30)$$

where

$$\phi(\mathcal{D}) = Hd(\mathcal{D}) + c_1(\mathcal{D}) + F_1(\mathcal{D}) \qquad (15.31)$$

(ii) *The backward equation* (15.27) *then becomes the optimality equation for the value function transform*

$$c(w, z) + \frac{\partial F(w, z)}{\partial t} + (w + z)F(w, z) - \phi(w)'H^{-1}\phi(z) = 0 \qquad (15.32)$$

(iii) *If F has an infinite-horizon limit under this optimal policy, with* ϕ *consequently*

196   CONTINUOUS TIME: THE PATH-INTEGRAL FORMALISM

having a time-independent limit value then

$$\Phi(z) = \overline{\phi(z)}H^{-1}\phi(z) \tag{15.33}$$

Proof. Relation (15.30), with ϕ defined by (15.31), follows from the stationarity condition of (15.26). This can be expressed as

$$\kappa(z) = H^{-1}(C_1(z) + F_1(z))$$

and relations (15.32) and (15.33) follow from this substitution for κ in (15.27) and (15.28). Note that F, and so F_1 and ϕ, may be t-dependent in (15.30)–(15.32). In (15.33) ϕ has its infinite-horizon limit value. □

Relation (15.33) is the continuous-time analogue of (12.31) and, like that, states the significant conclusion.

Corollary 15.3.4. *Relation (15.33) constitutes a canonical factorization of Φ, with ϕ related to the infinite-horizon value function transform F by (15.31). It satisfies the normalization that both Φ_0^{-1} and the coefficient of the highest powers in $\Phi_+ = \overline{\Phi}_-$ equal the known symmetric matrix H.*

The factorization (15.33) has been established, and its canonical nature follows, as always, from the presumed stable character of the 'optimally controlled plant equation' (15.30) if infinite-horizon limits exist.

15.3.1 Exercises and comments

(1) Consider the first-order case, for which we can write c in symmetric matrix form as

$$c(\zeta) = \begin{bmatrix} \zeta \\ \mathscr{D}\zeta \end{bmatrix}' \begin{bmatrix} c_{00} & c_{01} \\ c_{10} & c_{11} \end{bmatrix} \begin{bmatrix} \zeta \\ \mathscr{D}\zeta \end{bmatrix}$$

Then $F(\zeta) = \zeta'\Pi\zeta$ and $H = c_{11}$, $c_1(z) = c_{10}$, $F(w, z) = F_1(z) = \Pi$. The canonical factorization (15.33) becomes

$$c_{00} + c_{01}z - c_{10}z - c_{11}z^2 = (\Pi + c_{01} - c_{11}z)c_{11}^{-1}(\Pi + c_{10} + c_{11}z)$$

This can be reduced to what is effectively an equilibrium Riccati equation for Π:

$$c_{00} = (\Pi + c_{01})c_{11}^{-1}(\Pi + c_{10})$$

Of course, all these first-order formulae can be derived directly; the generating-function formalism is unnecessary in this case.

(2) Factorization (15.33) can itself be seen as an equilibrium Riccati equation for $F_1(z)$. It can be rewritten

$$\Phi - \bar{c}_1 d - \bar{d}c_1 - \bar{d}Hd = \bar{F}_1 d + \bar{d}F_1 + \overline{(c_1 + F_1)}H^{-1}(c_1 + F_1)$$

In the left-hand member we recognize $\Phi(z)$ stripped of all its highest-order terms.

15.4 THE PI/NR ALGORITHM

The successive approximations $\phi_{(i)}$ to the canonical factor ϕ in (15.33) generated by the Newton–Raphson algorithm will satisfy

$$\bar{\phi}_{(i+1)}H^{-1}\phi_{(i)} + \bar{\phi}_{(i)}H^{-1}\phi_{(i+1)} = \Phi + \bar{\phi}_{(i)}H^{-1}\phi_{(i)} \tag{15.34}$$

This is the continuous-time analogue of relation (12.38), simpler in that the central factor H^{-1} is fixed and known and need not be updated.

As in Section 12.4, recursion (15.34) is also that which would be generated by the policy-improvement algorithm. The line of proof will now be familiar; we outline it for this case in Exercise 15.1.1.

We have the analogue of Theorem 12.5.1 for stable $\phi_{(i)}$.

Theorem 15.4.1. *Equation (15.34) for $\phi_{(i+1)}$ has the solution*

$$\phi_{(i+1)} = H[\bar{\phi}_{(i)}^{-1}\Phi\phi_{(i)}^{-1}]_+\phi_{(i)} \tag{15.35}$$

and this recursion can be regarded as constituting the PI/NR algorithm.

Here the interpretation of the truncation operator $[\cdot]_+$ is that if a function $f(z)$ has an integral expression

$$f(z) = \int_{-\infty}^{\infty} e^{zt}dh(t) \tag{15.36}$$

valid on the imaginary z-axis then

$$[f(z)]_+ = \int_{-\infty}^{0+} e^{zt}dh(t) \tag{15.37}$$

This is the part of $f(z)$ which is analytic for $Re(z) \geq 0$.

Proof. Let us, as in the proof of Theorem 15.5.1, set $\phi_{(i+1)} = \phi$ and $\phi_{(i)} = \rho$ for simplicity. From equation (15.34) we deduce that

$$\bar{\rho}^{-1}\bar{\phi}H^{-1} + H^{-1}\phi\rho^{-1} = \bar{\rho}^{-1}\Phi\rho^{-1} + H^{-1} \tag{15.38}$$

Applying the truncation operator we deduce that

$$H^{-1}\phi\rho^{-1} = [\bar{\rho}^{-1}\Phi\rho^{-1}]_+ \tag{15.39}$$

whence (15.35) follows.

We have appealed to two facts in passing from (15.38) to (15.39). One is that, since ϕ and ρ are polynomial (in z) and ρ is stable, then $\phi\rho^{-1}$ is a rational function with all its singularities in the left half-plane. Thus, in an integral representation (15.36) of $\phi\rho^{-1}$ all contribution would come from $t \leq 0$, and we can assert that $[\phi\rho^{-1}]_+ = \phi\rho^{-1}$.

Correspondingly, all contributions to the integral representation of $\bar{\rho}^{-1}\bar{\phi} - I$

come from $t \geq 0$. In order to assert that $[\bar{\rho}^{-1}\bar{\phi} - \mathrm{I}]_+ = 0$ we would have to establish that no contribution comes from $t = 0$. However, $\bar{\rho}$ and $\bar{\phi}$ have the same highest-order term $\bar{d}H$. This will cancel in the numerator of $\bar{\rho}^{-1}\bar{\phi} - \mathrm{I}$, so that $\bar{\rho}^{-1}\bar{\phi} - \mathrm{I} = \mathrm{O}(z^{-1})$ for large $|z|$, whence all conclusions follow. □

15.4.1 Exercises and comments

(1) We give a sketch proof that the recursion (15.34) would also follow from policy improvement. If the choice of factor $\phi_{(i)}$ corresponds to the policy (15.23) then

$$\phi_{(i)} = H(d - \kappa) \tag{15.40}$$

By (15.31) an improved policy must correspond to

$$\phi_{(i+1)} = Hd + c_1 + F_1 \tag{15.41}$$

Here F_1 is derived from the infinite-horizon value function F under policy (15.40), and satisfies equation (15.28). Eliminating κ and F_1 from relations (15.40), (15.41) and (15.28), we deduce (15.34).

(2) Verify that the normalization asserted in Corollary 15.3.4 is conserved under recursion (15.34).

CHAPTER 16

Continuous time: control optimization

As for the discrete-time case in Chapter 13, we now transfer the general path-integral formalism to the control-optimization context.

16.1 THE MARKOV CASE REVISITED

Let us now treat the Markov regulation problem of Section 15.1 by the formalism of sections 15.2 and 15.3. As ever, we take $\zeta = (x, u, \lambda)$, and it is now natural to take

$$\mathbf{c}(\zeta) = c(x, u) + 2\lambda'(\dot{x} - Ax - Bu) - \theta\lambda'N\lambda \tag{16.1}$$

(c.f. (15.1)). The stationarity conditions for the path integral are $\Phi(\mathcal{D})\zeta = 0$ with

$$\Phi(\mathcal{D}) = \begin{bmatrix} R & S' & -\mathcal{D} - A' \\ S & Q & -B \\ \mathcal{D} - A & -B & -\theta N \end{bmatrix} \tag{16.2}$$

(c.f. (15.2)). The highest-order derivatives occurring in $\mathbf{c}$ are $\dot{x}$, u and λ, and so

$$H = \begin{bmatrix} 0 & 0 & I \\ 0 & Q & -B' \\ I & -B & -\theta N \end{bmatrix} \tag{16.3}$$

The path-integral value function (i.e. the extremal infinite-horizon path integral from initial values (x, u, λ)) turns out to be simply $x'\Pi x$, so the optimality equation (15.26) in the general setting and in equilibrium becomes

$$\operatorname*{stat}_{\dot{x}, u, \lambda} [c(x, u) + 2\lambda'(\dot{x} - Ax - Bu) - \theta\lambda'N\lambda + 2x'\Pi\dot{x}] = 0 \tag{16.4}$$

The stationarity conditions are

$$\phi(\mathcal{D})\zeta = 0 \tag{16.5}$$

where

$$\phi(\mathcal{D}) = \begin{bmatrix} \Pi & 0 & I \\ S & Q & -B' \\ \mathcal{D} - A & -B & -\theta N \end{bmatrix} \tag{16.6}$$

By its derivation, this is just the $\phi(\mathcal{D})$ defined in (15.31), and the canonical factorization of Φ is then

$$\Phi = \bar{\phi} H^{-1} \phi \tag{16.7}$$

with H, ϕ given by (16.3), (16.6).

This differs from either of the factorizations given in Section 15.1. In particular, it differs from the factorization (15.7), (15.8) obtained as a limit version of the 'standard' discrete-time factorization. This is because the continuous-time formulation has effectively impelled one to modify the definition of the instantaneous cost $c(\zeta)$ by a reallocation of cost over neighbouring instants of time.

The factorization deduced above is easily transformed to either of the factorizations of Section 15.1; we give details for the general case in the next section. For the moment, we note only the form of H^{-1}:

$$H^{-1} = \begin{bmatrix} J & -BQ^{-1} & I \\ -Q^{-1}B' & Q^{-1} & 0 \\ I & 0 & 0 \end{bmatrix} \tag{16.8}$$

with the spontaneous appearance of $J = BQ^{-1}B' + \theta N$.

16.2 THE HIGHER-ORDER MODEL

As in Section 1.1.11, we assume a plant equation

$$\mathscr{A} x + \mathscr{B} u = \varepsilon \tag{16.9}$$

with $\mathscr{A} = \sum_r A_r \mathscr{D}^r$, $\mathscr{B} = \sum_r B_r \mathscr{D}^r$. Let r_j be the highest order of derivative of the component x_j of x occurring in the plant equation, and let q_k be the corresponding highest order for u_k. If A_r has jkth element $a_{jk}^{(r)}$ we then define

$$A_* = (a_{jk}^{(r_k)}) \tag{16.10}$$

as the matrix of coefficients of highest-order derivatives of x occurring in (16.9). Correspondingly, we define

$$B_* = (b_{jk}^{(q_k)}) \tag{16.11}$$

Lemma 16.2.1. *For the plant equation* (16.9) *to constitute a valid forward recursion for x it is necessary that A_* should be non-singular. The equation can be transformed to normalize A_* to the identity matrix.*

The assertion is obvious; the matrix A_* plays the same role as does the matrix A_0 of the discrete-time case.

We can also generalize the instantaneous cost function to the form

$$c(x, u) = \sum_r \sum_s \begin{bmatrix} x^{[r]} \\ u^{[r]} \end{bmatrix}' \begin{bmatrix} R^{(rs)} & (S^{(sr)})' \\ S^{(rs)} & Q^{(rs)} \end{bmatrix} \begin{bmatrix} x^{[s]} \\ u^{[s]} \end{bmatrix} \tag{16.12}$$

where $x^{[r]} = \mathscr{D}^r x$, etc. The path integral then becomes

$$\mathbb{I}(\zeta) = \int [c(x,u) + 2\lambda'(\mathscr{A}x + \mathscr{B}u) - \theta\lambda'N\lambda]\,d\tau + \text{end effects} \quad (16.13)$$

leading to the stationarity condition $\Phi(\mathscr{D})\zeta = 0$ with

$$\Phi = \begin{bmatrix} R & \bar{S} & \bar{\mathscr{A}} \\ S & Q & \bar{\mathscr{B}} \\ \mathscr{A} & \mathscr{B} & -\theta N \end{bmatrix} \quad (16.14)$$

Here the $\mathscr{D}$ argument is understood in Φ and its entries, and we have, for example,

$$R(\mathscr{D}) = \sum_r \sum_s R^{(rs)}(-\mathscr{D})^r \mathscr{D}^s \quad (16.15)$$

(c.f. (15.16)). The conjugate $\bar{R}$ of R is defined, as always in continuous time, as $R(-\mathscr{D})'$. We can assume that $\bar{R} = R$ and $\bar{Q} = Q$ as a matter of normalization, and so $\bar{\Phi} = \Phi$.

16.3 THE CANONICAL FACTORIZATION

We shall now consider what form the canonical factorization (15.33) takes for the control model of the previous section. We shall make the simplifying assumptions that $c(x, u)$, defined in (16.12), involves derivatives of x_j and of u_k of orders *less than* r_j and *up to* q_k, respectively, for all relevant j and k. Under these circumstances we have

$$H = \begin{bmatrix} 0 & 0 & A'_* \\ 0 & Q_* & B'_* \\ A_* & B_* & -\theta N \end{bmatrix} \quad (16.16)$$

where

$$Q_* = (Q_{jk}^{(q_j, q_k)}) \quad (16.17)$$

is the matrix of coefficients of the highest-order derivatives of u in $c(x, u)$. We shall assume that the derivatives of u_k occur in $c(x, u)$ up to order q_k strictly, in that Q_* is non-singular (and necessarily positive-definite, if the cost (16.12) is assumed non-negative).

These assumptions constitute a natural extension of the pattern of the first-order case, for which $r_j = 1$, $q_k = 0$ and $c(x, u)$ involves derivatives of zeroth order only. They would be inappropriate only if one had reason to penalize large values of derivatives of higher order than those indicated.

We shall derive the form of the canonical factor ϕ, not by substituting in formula (15.31), but by duplicating its derivation. That is, relation (15.30) is the stationarity condition for the optimality equation in the path-integral formulation.

Let $F(x, u)$ be the infinite-horizon value function for the control process; it is in fact a function of the derivatives $x_j^{[r]}$ ($0 \le r < r_j; j = 1, 2, \ldots, n$) and $u_k^{[s]}$ ($0 \le s < q_k; k = 1, 2, \ldots, m$). One verifies that this is also the infinite-horizon value function from initial (x, u, λ) in the path-integral formulation. The quantity to be extremized with respect to the highest-order derivatives of $\zeta = (x, u, \lambda)$ in the optimality equation is thus

$$c(x, u) + \lambda'(\mathscr{A}x + \mathscr{B}u) - \theta \lambda' N \lambda + \sum_j x_j^{[r_j]} \frac{\partial F}{\partial x_j^{[r_j - 1]}} + \sum_k u_k^{[q_k]} \frac{\partial F}{\partial u_k^{[q_k - 1]}} \quad (16.18)$$

with $c(x, u)$ given by expression (16.12). The F-derivatives will be linear in the derivatives of x and u; let us then write the vectors of these derivatives as

$$\left[\frac{\partial F(x, u)}{\partial x_j^{[r_j - 1]}} \right] = F_{11}(\mathscr{D})x + F_{12}(\mathscr{D})u$$
$$\left[\frac{\partial F(x, u)}{\partial u_j^{[q_j - 1]}} \right] = F_{21}(\mathscr{D})x + F_{22}(\mathscr{D})u \quad (16.19)$$

Correspondingly, let us define

$$\left[\frac{\partial c(x, u)}{\partial u_j^{[q_j]}} \right] = C_1(\mathscr{D})x + C_2(\mathscr{D})u \quad (16.20)$$

so that, in fact,

$$C_1(\mathscr{D}) = \left[\sum_s S_{jk}^{(q_j, s)} \mathscr{D}^s \right]$$
$$C_2(\mathscr{D}) = \left[\sum_s Q_{jk}^{(q_j, s)} \mathscr{D}^s \right] \quad (16.21)$$

Theorem 16.3.1. *Under the assumptions listed, the matrix (16.14) has the canonical factorization*

$$\Phi(z) = \overline{\phi(z)} H^{-1} \phi(z) \quad (16.22)$$

with H given by formula (16.16) and

$$\phi(z) = \begin{bmatrix} F_{11}(z) & F_{12}(z) & A'_* \\ F_{21}(z) + C_1(z) & F_{22}(z) + C_2(z) & B'_* \\ \mathscr{A}(z) & \mathscr{B}(z) & -\theta N \end{bmatrix} \quad (16.23)$$

Here the entries F_{jk} are expressed in terms of the value function of the control problem by (16.19) and C_1 and C_2 in terms of the instantaneous cost function by (16.21).

Proof. If we write down the stationarity conditions of the form (16.18) with respect to $x_j^{(r_j)}$ ($j = 1, 2, \ldots, n$), $u_k^{(q_k)}$ ($k = 1, 2, \ldots, m$) and λ (which are the highest-

16.3 THE CANONICAL FACTORIZATION

order derivatives of x, u and λ occurring) then we obtain an equation system $\phi(\mathscr{D})\zeta = 0$ with ϕ given by (16.23). □

Note a degree of simplification: the instantaneous cost function does not contribute to the first row of ϕ because it does not contain the derivatives $x_j^{[r_j]}$.

Corollary 16.3.2. *Assume A_* normalized to the identity. Then the relation determining the optimal control in closed-loop form is*

$$[C_1(\mathscr{D}) + F_{21}(\mathscr{D}) - B'_* F_{11}(\mathscr{D})]x + [C_2(\mathscr{D}) + F_{22}(\mathscr{D}) - B'_* F_{12}(\mathscr{D})]u = 0 \tag{16.24}$$

If u does not appear in undifferentiated form in either plant equation or instantaneous cost function (so that one can set $B_* = -B$ and $Q_* = Q$) then (16.24) reduces to

$$u = -Q^{-1}[S(\mathscr{D}) + B'F_{11}(\mathscr{D})]x \tag{16.25}$$

Proof. Relation (16.24) follows by elimination of λ between the two first relations of $\phi(\mathscr{D})\zeta = 0$. If u does not occur in differentiated form then F_{12}, F_{21} and F_{22} are all zero and $C_1(\mathscr{D}) = S(\mathscr{D})$, $C_2(\mathscr{D}) = Q$. □

Relation (16.24) is to be regarded as a differential equation for u (of order q_k in u_k) driven by known derivatives of x (of order less than r_j in x_j). Relation (16.25) generalizes the familiar $u = -Q^{-1}(S + B'\Pi)x$.

Factorization (16.22) will imply a set of Riccati equations for the quantities $F_{jk}(z)$. The fact that only these occur rather than the full value function transform $F(w, z)$ is a consequence of the fact that, as in Chapter 12, the former determine the latter. These Riccati equations actually take the simplest form if we transform the canonical factor to

$$\begin{bmatrix} I & 0 & 0 \\ -B'_* & I & 0 \\ 0 & 0 & I \end{bmatrix} \phi = \begin{bmatrix} \mathfrak{P} & \mathfrak{A}'_* \\ \mathfrak{A} & -\theta N \end{bmatrix} = \begin{bmatrix} P_{11} & P_{12} & I \\ P_{21} & P_{22} & 0 \\ \mathscr{A} & \mathscr{B} & -\theta N \end{bmatrix}$$

where

$$\left.\begin{array}{l} P_{1k} = F_{1k} \\ P_{2k} = C_k + F_{2k} - B'_* F_{1k} \end{array}\right\} \quad (k = 1, 2) \tag{16.26}$$

These latter are just the matrices occurring in the optimal control relationship (16.24). In terms of these transformed factors the factorization (16.22) becomes

$$\Phi = \begin{bmatrix} \bar{\mathfrak{P}} & \bar{\mathfrak{A}} \\ \mathfrak{A}_* & -\theta N \end{bmatrix} \begin{bmatrix} \theta N & 0 & I \\ 0 & Q_*^{-1} & 0 \\ I & 0 & 0 \end{bmatrix} \begin{bmatrix} \mathfrak{P} & \mathfrak{A}'_* \\ \mathfrak{A} & -\theta N \end{bmatrix} \tag{16.27}$$

whence we deduce:

Corollary 16.3.3. *Factorization* (16.27) *implies the Riccati equations for the* P_{jk}:

$$\mathfrak{R} = \bar{\mathfrak{P}}\begin{bmatrix} \theta N & 0 \\ 0 & Q_*^{-1} \end{bmatrix}\mathfrak{P} + \bar{\mathfrak{P}}\begin{bmatrix} \mathfrak{A} \\ 0 \end{bmatrix} + [\bar{\mathfrak{A}} \ 0]\mathfrak{P} \qquad (16.28)$$

or, in full,

$$R = \theta \bar{P}_{11} N P_{11} + \bar{P}_{21} Q_*^{-1} P_{21} + \bar{P}_{11}\mathscr{A} + \bar{\mathscr{A}} P_{11}$$
$$S = \theta \bar{P}_{12} N P_{11} + \bar{P}_{22} Q_*^{-1} P_{21} + \bar{P}_{12}\mathscr{A} + \bar{\mathscr{B}} P_{11} \qquad (16.29)$$
$$Q = \theta \bar{P}_{12} N P_{12} + \bar{P}_{22} Q_*^{-1} P_{22} + \bar{P}_{12}\mathscr{B} + \bar{\mathscr{B}} P_{12}$$

It may seem that (16.29) provides only three relationships to determine four functions. However, the second relation counts double, as the equations implied by equating coefficients of positive and of negative powers of z to zero are distinct.

In the case when u occurs only in undifferentiated form then P_{22} reduces simply to $Q_* = Q$, and we see from (16.24) that P_{21} must be interpretable as $QK(z)$, where $u = K(\mathscr{D})x$ is the optimal control relationship. Relations (16.29) then amount to

$$R = \theta \bar{P}_{11} N P_{11} + \bar{K} Q K + \bar{P}_{11}\mathscr{A} + \bar{\mathscr{A}} P_{11}$$
$$S = QK - B'P_{11}$$

and an identity. Recalling that $P_{11} = F_{11}$, we then have

Corollary 16.3.4. *If* u *occurs in the problem only in undifferentiated form then the optimal control relation is given by* (16.25), *where* F_{11} *is determined by the Riccati equation*

$$R^* = \bar{F}_{11}\mathscr{A}^* + \bar{\mathscr{A}}^*F_{11} + \bar{F}_{11}(BQ^{-1}B' + \theta N)F_{11} \qquad (16.30)$$

Here, as ever, $\mathscr{A}^* = \mathscr{A} - \mathscr{B}Q^{-1}S$ and $R^* = R - \bar{S}Q^{-1}S$ are the values obtained under normalization of S to zero.

The most useful exercise would now be the numerical treatment of a number of control examples by application of the PI/NR algorithm, (15.34) or (15.35), exploiting the simplification that the canonical factors are known to take the special form indicated in Theorem 16.3.1. It is hoped that our analysis has produced a sufficient reduction that others may see this project as worthwhile.

16.3.1 Exercises and comments

(1) Consider a scalar case with zero S, constant R and Q, and plant equation $\dot{x} = Bu + \varepsilon$. Show, from (16.25) and (16.30), that the optimal infinite-horizon control is $u = -Q^{-1}B(R/J)^{1/2}x$, where $J = B^2Q^{-1} + \theta N$, as ever.

(2) Suppose Exercise (1) varied only in that the plant equation is $\ddot{x} = Bu + e$. Show that the optimal control is $u = -Q^{-1}B[(R/J)^{1/2}x + (4R/J^3)^{1/4}\dot{x}]$.

PART IV

Connections and variations

CHAPTER 17

The relationship of the LEQG criterion to H_∞ and entropy criteria

17.1 NOTATION

For much of this chapter we shall find it very convenient to use the system notation introduced in Section 10.7. Consider the homogeneous problem of regulation to zero, with only white-noise disturbance. We assume the current plant variable observable. The system notation of Section 10.7 will then lead us to define the quantities

$$\mathfrak{x} = \begin{bmatrix} x \\ u \end{bmatrix} \quad \mathfrak{A} = [\mathscr{A} \; \mathscr{B}] \quad \mathfrak{R} = \begin{bmatrix} R & S' \\ S & Q \end{bmatrix}$$

in terms of which the plant equation and the instantaneous cost function become

$$\mathfrak{A}\mathfrak{x} = \varepsilon \tag{17.1}$$

$$c(\mathfrak{x}) = \mathfrak{x}'\mathfrak{R}\mathfrak{x} \tag{17.2}$$

We shall consider only the discrete-time case, so that $\mathfrak{A}$ is a distributed-lag operator of the form

$$\mathfrak{A} = \mathfrak{A}(\mathscr{T}) = \sum_{r=0}^{p} \mathfrak{A}_r \mathscr{T}^r$$

However, we shall assume $\mathfrak{R}$ constant, so that $c(\mathfrak{x})$ is a function of current $\mathfrak{x}$ alone. This is not necessary, but eases comparison of the LEQG formulation with the standard form of others.

We shall use generating functions extensively. Any expansion

$$f(z) = \sum_{j=-\infty}^{\infty} f_j z^j \tag{17.3}$$

is understood as that which is valid on the unit circle, unless otherwise stated. The absolute term f_0 in such an expansion will be denoted $\text{Abs} f(z)$. Thus

$$\text{Abs} f(z) = \frac{1}{2\pi i} \oint f(z) \frac{dz}{z} = \frac{1}{2\pi} \int_{-\pi}^{\pi} f(e^{-i\omega}) d\omega \tag{17.4}$$

where the first integral is around the unit circle.

The coefficients f_j are, in general, real matrices. The conjugate of $f(z)$ is then

$$\overline{f(z)} = \sum_j f_j' z^{-j} \tag{17.5}$$

We shall use $[f]_+$ to denote

$$[f(z)]_+ = \sum_{j=0}^{\infty} f_j z^j \tag{17.6}$$

the truncation of expansion (17.3) to non-negative powers. We shall also use 'neg' to denote a function of z whose power series expansion on $|z| = 1$ includes only negative powers of z. Thus

$$f = [f]_+ + \text{neg}$$

17.2 LQG AND H_∞ CONTROL CRITERIA

We assume that $\mathfrak{R}$ and $N = \text{cov}(\varepsilon)$ are non-negative-definite, and can thus be factorized:

$$\mathfrak{R} = D'D \qquad N = WW' \tag{17.7}$$

where D and W are square. The plant equation (17.1) and cost function (17.2) can then be written

$$\mathfrak{A}x = Wv \tag{17.8}$$

$$c(x) = \chi'\chi = |\chi|^2 \tag{17.9}$$

where v is 'standard' white noise (i.e. $\text{cov}(v) = I$) and

$$\chi = Dx \tag{17.10}$$

We can regard χ as the vector of 'deviations', in that the control optimization problem has been framed as the minimization of $|\chi|$ in some stochastic sense.

In this chapter we shall concentrate on the optimization of stationary control rules, and so must assume that stabilizing control rules exist, which make $\{x_t\}$ a stationary process. More specifically, under such a policy x_t will be representable as a time-homogeneous linear function of input

$$x_t = \sum_{j=0}^{\infty} H_j v_{t-j} \tag{17.11}$$

and the $v \to x$ *transfer function* $H(z) = \sum_j H_j z^j$ will be analytic in $|z| \leq 1$. The transfer function $v \to \chi$ will be $G(z) = DH(z)$.

Since the aim is to make χ as small as possible in some sense, then, in the stationary regime, the aim must be to make the transfer function $G(z)$ as small as possible in some sense. Different criteria take different measures of size. The LQG criterion (in the stationary regime) is to choose a control rule which minimizes the average cost

$$Ec(x) = E[x'\mathfrak{R}x] = E[|\chi|^2] = \text{Abs Tr}[\overline{G(z)}G(z)] \tag{17.12}$$

where Tr denotes the operation of taking the trace of a matrix. This expectation can alternatively be written

$$Ec(\mathbf{x}) = \operatorname{Abs}\operatorname{Tr}[\overline{H(z)}D'DH(z)] = \operatorname{Abs}\operatorname{Tr}[\overline{H(z)}\Re H(z)] \quad (17.13)$$

but it is expression (17.12) that we wish to stress at this moment.

The H_∞ criterion is that the control shall be chosen so as to minimize the maximum singular value of the transfer function $G(z)$ on $|z| = 1$. (The maximum, that is, over all its singular values and over the value of z.) This is equivalent to minimizing the maximum eigenvalue of $\overline{G(z)}G(z)$ on $|z| = 1$.

Since $G(e^{-i\omega})$ is the frequency-response function of the output χ to input ν the H_∞ criterion optimizes 'worst behaviour' in frequency response. For accounts of H_∞-theory see, for example, Francis (1987) or Zames (1981).

We see from the above discussion that the LQG and H_∞ criteria can be respectively regarded as average and minimax criteria based on the cost-function $|\chi|^2$. As is familiar from other contexts, the analyses based upon these two extreme criteria show some analogy. Indeed, we shall see in the next section that they are particular members of a family of criteria which are effectively the LEQG criteria for varying θ.

Lemma 17.2.1. *The maximal eigenvalue of* $\overline{G(z)}G(z)$ *can be written* $-\bar{\theta}^{-1}$, *where* $\bar{\theta}$ *(necessarily negative) is the maximum on* $|z| = 1$ *of the maximal θ-root of*

$$|I + \theta \bar{G} G| = 0 \quad (17.14)$$

This is equivalently the maximal root of

$$|I + \theta \bar{H} \Re H| = 0 \quad (17.15)$$

and an H_∞-optimal control minimizes $\bar{\theta}$.

This lemma is just a convenient summary of evident equivalences. The maximal eigenvalue of $\overline{G(z)}G(z)$ for given z is just $-\theta(z)^{-1}$, where $\theta(z)$ is the maximal root of (17.14) for the given z. All equivalences then follow. Note that we have left the z-argument to be understood in (17.14) and (17.15); a convenience which seldom causes confusion.

17.3 ENTROPY AND LEQG CONTROL CRITERIA

A control criterion associated with the H_∞ concept is that of choosing the control rule to minimize the expression

$$\mathscr{E}(\theta) = \theta^{-1} \operatorname{Abs} \log |I + \theta \bar{G} G| = \frac{1}{2\pi\theta} \int_{-\pi}^{\pi} \log |I + \theta \overline{G(e^{-i\omega})} G(e^{-i\omega})| \, d\omega$$

$$= \frac{1}{2\pi\theta} \int_{-\pi}^{\pi} \log |I + \theta \bar{H} \Re H| \, d\omega \quad (17.16)$$

This is the *minimal entropy* criterion. Its motivation was unclear for some time, except that it also measures the size of the frequency transfer function $G(e^{-i\omega})$ if $\theta > \bar{\theta}$, and that it seems to arise naturally if one considers the H_∞ criterion, expressed in terms of the maximal root of (17.14).

The minimal entropy criterion will make sense only if $\theta > \bar{\theta}$. As $\theta \downarrow \bar{\theta}$ the class of entropy-minimizing rules converges to just the class of H_∞-optimal rules. All these matters are much clarified by the following theorem, due to Glover and Doyle (1988).

Theorem 17.3.1. *The minimum-entropy criterion is equivalent to the average-optimal form of the LEQG criterion, with θ identified as the risk-sensitivity parameter.*

By the 'average-optimal' form of the LEQG criterion we mean that we seek a π minimizing the limiting value

$$\delta_\pi(\theta) = \lim_{h \to \infty} h^{-1} \gamma_\pi(\theta) = \lim_{h \to \infty} \left[-(2/\theta h) \log E_\pi \exp\left(-(\theta/2) \sum_1^h |\chi_t|^2 \right) \right] \quad (17.17)$$

This limit will certainly exist for any stationary stabilizing policy π. If the π minimizing $\gamma_\pi(\theta)$ has a stationary limit form for $h \to \infty$ then this limit is average-optimal, although the converse is not necessarily true (see e.g. Blackwell, 1962).

The equivalence asserted by Theorem 17.3.1 means that the LEQG and H_∞ formulations, apparently quite different, are in fact very close. Our extensions of LQG theory to the LEQG case then supply a theory for the minimal-entropy and H_∞ versions (in their limiting stationary forms). The recursive techniques which are so useful for finite-order systems will now be replaced by analytic techniques much more akin to those employed in Part III.

Proof of the Glover–Doyle theorem is best preceded by a few lemmas. Let us define the autocovariance matrix

$$V_s = E(\chi_t \chi'_{t-s}) \quad (17.18)$$

of the χ-process.

Lemma 17.3.2.
$$E \exp(-(\theta/2)\chi'\chi) = |I + \theta V_0|^{-1/2} \quad (17.19)$$

This follows by the evaluation of a standard multivariate normal integral.

Lemma 17.3.3. *The criterion $\gamma_\pi(\theta)$ of (17.17) has the evaluation*

$$\gamma_\pi(\theta) = \theta^{-1} \log |\Sigma_h| \quad (17.20)$$

where Σ_h is the partitioned matrix with jkth matrix component

$$(\Sigma_h)_{jk} = \delta_{jk} I + \theta V_{j-k} \quad (j, k = 1, 2, \ldots, h) \quad (17.21)$$

17.3 ENTROPY AND LEQG CONTROL CRITERIA

This follows from an application of Lemma 17.3.2, with χ replaced by the column vector with vector components $\chi_1, \chi_2, \ldots, \chi_h$.

Denote expression (17.21) by σ_{j-k} and define the generating function

$$\sigma(z) = \sum_{-\infty}^{\infty} \sigma_j z^j \qquad (17.22)$$

This is Hermitian, and will be positive on the unit circle if convergent and if θ does not take too negative a value.

Lemma 17.3.4. *Suppose $\sigma(z)$ absolutely convergent on the unit circle. Then*

$$\lim_{h \to \infty} h^{-1} \log |\Sigma_h| = \text{Abs} \log |\sigma(z)| \qquad (17.23)$$

Results such as this were derived heuristically in Whittle (1951, 1953). One approach is indicated in Exercise 17.3.1. For rigorous treatments see Grenander and Szegö (1958) and Hannan (1970).

Applying Lemma 17.3.4 to the evaluation of expression (17.20) we deduce that

$$\delta_\pi(\theta) = \lim_{h \to \infty} \gamma_\pi(\theta) = \theta^{-1} \text{Abs} \log |I + \theta \bar{G} G| = \mathscr{E}(\theta) \qquad (17.24)$$

This proves Theorem 17.3.1, in that it indicates the asymptotic identity of the LEQG criteron $h^{-1}\gamma_\pi(\theta)$ and the entropy expression (17.16).

17.3.1 Exercise and comments

(1) We know that $|\Sigma_h|$ is the product of the eigenvalues of Σ_h, i.e. of the rh values of the scalar α such that

$$\sum_k \sigma_{j-k} \xi_k = \alpha \xi_j \qquad (17.25)$$

Here r is the dimension of σ, the ξ_j are r-vectors, and j and k take values $1, 2, \ldots, h$. Assume now that σ_j is periodic in j with period h. Then

$$\xi_j = e^{2\pi i s/h} \xi \qquad (17.26)$$

with s integral solves (17.25) if α and ξ solve the r-dimensional eigenvalue problem

$$\sigma(e^{2\pi i s/h}) \xi = \alpha \xi \qquad (17.27)$$

The product of the eigenvalues of the system (17.27) is the determinant $|\sigma(e^{2\pi i s/h})|$. We account for all rh eigenvalues of the system (17.25) by taking $s = 1, 2, \ldots, h$, and obtain

$$h^{-1} \log |\Sigma_h| = h^{-1} \sum_{s=1}^{h} \log |\sigma(e^{2\pi i s/h})| \qquad (17.28)$$

Equation (17.23) then follows as the limit form of (17.28). The convergence condition placed on $\sigma(z)$ in Lemma 17.3.4 is designed to ensure that the periodicity assumption is justifiable in the limit of large h.

17.4 THE SCALAR MARKOV CASE

The critical value $\bar{\theta}$ should be just the infinite-horizon value of the breakdown value defined in Section 6.4. We can check this by comparing the evaluation of $\bar{\theta}$ with that already made in Section 9.1 for the scalar, Markov state-observed case, with $S = 0$. The check is actually of some interest, because $\bar{\theta}$ has a different analytic evaluation in different parameter ranges, and it is useful to see how this comes about in the H_∞ formulation.

Suppose that a control exists which is stabilizing; i.e. we can choose a control $u = Kx$ such that $\Gamma = A + BK$ is less than unity in modulus. We can always normalize to the case $A > 0$ (by considering a series $(-)^t x_t$, if necessary) and the optimal Γ will then also be non-negative. We have then

$$H(z) = \begin{bmatrix} (1 - \Gamma z)^{-1} W \\ K(1 - \Gamma z)^{-1} W \end{bmatrix} \tag{17.29}$$

so that

$$|I + \theta \bar{H} \Re H| = \begin{vmatrix} 1 + \theta \beta R & \theta \beta R K \\ \theta \beta R K & 1 + \theta \beta Q K^2 \end{vmatrix} = 1 + \theta \beta (R + QK^2) \tag{17.30}$$

where

$$\beta = \frac{N}{(1 - \Gamma z)(1 - \Gamma z^{-1})}$$

The H_∞-optimal control will be that which minimizes the maximum on $|z| = 1$ of $\beta(R + QK^2)$, and we shall have

$$\bar{\theta} = -\left[\min_K \max_z \beta(R + QK^2) \right]^{-1} \tag{17.31}$$

where the maximum is over z on the unit circle. If $0 < \Gamma < 1$ then this maximum is at $z = 1$ and the H_∞-optimal control chooses K to minimize

$$\frac{R + QK^2}{(1 - A - BK)^2} \tag{17.32}$$

under the conditions

$$0 \leq A + BK < 1 \tag{17.33}$$

Theorem 17.4.1. *The critical value $\bar{\theta}$ under the H_∞ criterion is indeed just that identified for the LEQG case in Section 9.1, namely*:

$$\bar{\theta} = \theta_2 = -\frac{1}{N}\left[\frac{B^2}{Q} + \frac{(1 - |A|)^2}{R} \right] \tag{17.34}$$

if

$$A^2 - |A| \leq RB^2/Q \tag{17.35}$$

and

$$\bar{\theta} = \theta_3 = -\frac{1}{N}[R + QA^2/B^2]^{-1} \qquad (17.36)$$

otherwise.

Proof. Expression (17.32) attains its free minimum at

$$K = -\frac{BR}{Q(1 - A)} \qquad (17.37)$$

which corresponds to the critical value (17.34). However, this minimizing value is acceptable only if (17.33) holds, i.e. if

$$0 \leq A - \frac{B^2 R}{Q(1 - A)} < 1 \qquad (17.38)$$

The second inequality of (17.38) is always violated if $A \geq 1$. If we assume that $0 \leq A < 1$ then the first inequality of (17.38) is equivalent to (17.35). Since inequality (17.35) is also violated if $A \geq 1$, then (17.38) is violated if (17.35) is violated. When inequality (17.35) is violated then expression (17.32) is minimal subject to (17.33) for $\Gamma = A + BK = 0$; this corresponds to the critical value (17.36). □

17.5 LEQG-OPTIMAL CONTROL IN TRANSFORM TERMS

With the establishment in Theorem 17.3.1 of the equivalence (in the average sense) between the LEQG and entropy criteria there may seem to be nothing more to say. We have devoted Chapters 6–16 to the derivation of LEQG-optimal control rules, and these are now seen to be entropy-minimizing as well.

However, there is interest in checking directly that the LEQG-optimal rules are entropy-minimizing, without appeal to the Glover–Doyle theorem. First, this avoids some rather circuitous arguments, and, second, it is interesting that the methods are so different to those we have used previously. Our previous approaches appealed either to the recursive techniques of dynamic programming or (thanks to the risk-sensitive certainty-equivalence principle) to the extremization of a path integral. The approach will now be the determination of optimal filters via the determination of their transfer functions. The analytical steps which must be followed must somehow contain in transformed form the essentials of our previous work; in particular, the derivation of a risk-sensitive certainty-equivalence principle.

For simplicity, we shall discuss only the case set out in Section 17.1; of regulation to zero with a completely observed process variable, with higher-order effects in the plant equation but not in the cost function. We shall switch freely between operators and corresponding generating functions in that, for example, we shall use $\mathfrak{A}$ to denote both the operator $\mathfrak{A}(\mathcal{T})$ and the generating function

$\mathfrak{A}(z) = z^{-t}\mathfrak{A}(\mathcal{T})z^t$. The context will make clear which sense is current. In the plant equation (17.8) it is used in the first sense. However, if $H(z)$ is the transfer function $v \to x$ then this plant equation would imply that

$$\mathfrak{A}H = W \tag{17.39}$$

a relation in which $\mathfrak{A}$ is understood in the second sense.

In terms of H the entropy criterion (17.16) can be written

$$\mathscr{E}(\theta) = \theta^{-1}\operatorname{Abs}\log|I + \theta\bar{H}\Re H| \tag{17.40}$$

The transfer function H is to be chosen to minimize criterion (17.40) subject to the plant constraint (17.39) and subject also to realizability constraints. If the current process variable x is observable then control u_t can be an arbitrary function of v_{t-j} ($j \geq 0$). That is, $H(z)$ is constrained to the one-sided form

$$H(z) = \sum_{j=0}^{\infty} H_j z^j \tag{17.41}$$

valid as an expansion on $|z| = 1$. Let us now see how the LEQG-optimal control is to be characterized in these terms.

We shall, as ever, write

$$\Phi(z) = \begin{bmatrix} \Re & \bar{\mathfrak{A}} \\ \mathfrak{A} & -\theta N \end{bmatrix} \tag{17.42}$$

(c.f. Sections 10.2 and 10.7) with the canonical factorization in the standardized form

$$\Phi(z) = \overline{\phi(z)}\phi_0^{-1}\phi(z) \tag{17.43}$$

(c.f. Section 12.3). We know (Theorem 13.1.2) that this canonical factor can be written

$$\phi(z) = \begin{bmatrix} \mathfrak{D}(z) & \mathfrak{A}'_0 \\ \mathfrak{A}(z) & -\theta N \end{bmatrix} \tag{17.44}$$

Theorem 17.5.1. *The LEQG-optimal value of $H(z)$ obeys and is determined by the equation system*

$$\begin{bmatrix} \mathfrak{D}(z) & \mathfrak{A}'_0 \\ \mathfrak{A}(z) & 0 \end{bmatrix}\begin{bmatrix} H(z) \\ \lambda(z) \end{bmatrix} = \begin{bmatrix} 0 \\ W \end{bmatrix} \tag{17.45}$$

where, as in (17.41), H and λ have power series expansions in non-negative powers of z on $|z| = 1$.

Proof. We know from Section 12.3 that the equation system

$$\phi(\mathcal{T})\begin{bmatrix} x \\ \lambda \end{bmatrix} = 0 \tag{17.46}$$

17.6 DERIVATION OF THE OPTIMAL RISK-NEUTRAL CONTROL

determines the predicted course of the optimally controlled process. The first relation

$$\mathfrak{D}(\mathcal{T})\mathfrak{x} + \mathfrak{A}'_0 \lambda = 0 \qquad (17.47)$$

determines the optimal control rule, in that we know (c.f. equations (13.10) and (13.12)) that λ can be eliminated from it to determine u optimally in terms of current observables. The final relation

$$\mathfrak{A}(\mathcal{T})\mathfrak{x} - \theta N \lambda = 0$$

determines the *predicted* course of the optimally controlled process. The *actual* dynamic relation will be the plant equation

$$\mathfrak{A}(\mathcal{T})\mathfrak{x} = \varepsilon = Wv \qquad (17.48)$$

Relations (17.47) and (17.48) between them determine the actual course of the optimally controlled process. The variable λ at time t should then be seen as $\lambda_t^{(t)}$ in our earlier notation. In transfer function terms this pair of relations becomes exactly (17.45), with $\lambda(z)$ being the transfer function $v_t \to \lambda_t^{(t)}$. □

The formal variable λ can in fact be eliminated, and equation (17.45) solved for H (see Exercise 17.5.1). However, the analysis continues much more naturally if we do not make this reduction.

The control rule of course remains θ-dependent, because $\mathfrak{D}$ is determined from the canonical factorization of the matrix (17.42).

17.5.1 Exercise and comments

(1) If we write expression (17.44) in the more explicit form

$$\phi = \begin{bmatrix} D_{11} & D_{12} & I \\ D_{21} & D_{22} & 0 \\ \mathcal{A} & \mathcal{B} & -\theta N \end{bmatrix}$$

(the z-argument being taken as understood) then we see that λ can be eliminated from the system (17.45) to yield the solution

$$H = \begin{bmatrix} D_{21} & D_{22} \\ \mathcal{A} & \mathcal{B} \end{bmatrix}^{-1} \begin{bmatrix} 0 \\ W \end{bmatrix}$$

for the optimal transfer function.

17.6 DERIVATION OF THE OPTIMAL RISK-NEUTRAL CONTROL

It is helpful to begin with the risk-neutral case, for which the entropy criterion (17.40) reduces to

$$\mathscr{E}(0) = \operatorname{Abs} \operatorname{Tr}(\bar{H}\mathfrak{R}H) \qquad (17.49)$$

as we have already observed in (17.13). This is to be minimized subject to the plant and realizability constraints (17.39) and (17.40).

Set $\mathfrak{A}H - W = \sum_0^\infty M_j z^j$. Then the constraints $M_j = 0$ ($j \geq 0$) expressed by (17.39) could be accounted for by adding the Lagrangian combination

$$\text{Tr}\left[\sum_j \Lambda'_j M_j\right] = \text{Abs Tr}[\overline{\Lambda(z)}(\mathfrak{A}H - W)] = \text{Abs Tr}[\overline{(\mathfrak{A}H - W)}\Lambda(z)]$$

to the criterion (17.49). Here the elements of Λ_j are Lagrangian multipliers associated with the constraints implied by $M_j = 0$, and $\Lambda(z) = \sum_0^\infty \Lambda_j z^j$. In fact, we shall find it convenient to take $2\Lambda_j$ as the Lagrangian multiplier, so that the Lagrangian form becomes

$$L(H) = \text{Abs Tr}[\overline{H}\mathfrak{R}H + 2\overline{(\mathfrak{A}H - W)}\Lambda] \qquad (17.50)$$

Lemma 17.6.1. *The Lagrangian optimality condition for H is*

$$\mathfrak{R}H + \overline{\mathfrak{A}}\Lambda = \text{neg} \qquad (17.51)$$

(Recall the 'neg' convention defined in Section 17.1.)

Proof. If we modify H infinitesimally to $H + \Delta = \sum_0^\infty (H_j + \Delta_j) z^j$ then the stationarity condition for $L(H)$ becomes

$$\text{Abs Tr}[\overline{\Delta}\mathfrak{R}H + \overline{\Delta}\overline{\mathfrak{A}}\Lambda] = 0 \qquad (17.52)$$

Since Δ_j is arbitrary, then relation (17.52) implies that the term in z^j in $\mathfrak{R}H + \overline{\mathfrak{A}}\Lambda$ is zero. Since this holds for $j = 0, 1, 2, \ldots$ we derive relation (17.51). □

Theorem 17.6.2. *In the risk-neutral case $\theta = 0$ the transfer function H determined by (17.45) satisfies the LQG optimality condition (17.51).*

Proof. Relations (17.39) and (17.51) between them imply that

$$\Phi\begin{bmatrix} H \\ \Lambda \end{bmatrix} = \begin{bmatrix} \text{neg} \\ W \end{bmatrix} \qquad (17.53)$$

Here Φ is defined as in (17.42), although now with $\theta = 0$. Now

$$\overline{\phi}\phi_0^{-1} = I + \text{neg}$$

and the same will hold for the inverse of this matrix, so we deduce from (17.53) that

$$\phi\begin{bmatrix} H \\ \Lambda \end{bmatrix} = \begin{bmatrix} 0 \\ W \end{bmatrix} + \text{neg} \qquad (17.54)$$

However, since the left-hand member of (17.54) contains no 'neg' terms in its

17.7 DERIVATION OF THE OPTIMAL RISK-SENSITIVE CONTROL

expansion, this relation must reduce to

$$\phi \begin{bmatrix} H \\ \Lambda \end{bmatrix} = \begin{bmatrix} 0 \\ W \end{bmatrix} \tag{17.55}$$

However, this is exactly relation (17.43) (in the case $\theta = 0$) with the two generating functions of Lagrangian multipliers, $\lambda(z)$ and $\Lambda(z)$, identified. By Theorem 17.5.1, the H determined by (17.51) agrees with our earlier determination of the optimal stationary LQG rule. □

It is probably the statement that the 'neg' term in (17.57) is zero, which amounts to statement of a certainty-equivalence principle. The corresponding risk-sensitive version of this argument is more complex.

17.7 DERIVATION OF THE OPTIMAL RISK-SENSITIVE CONTROL

We now consider the entropy criterion (17.40) for arbitrary $\theta > \bar{\theta}$. We shall account for the plant constraint by considering the Lagrangian form

$$L(H) = \mathcal{E}(\theta) + 2 \operatorname{Abs} \operatorname{Tr}[\overline{(\mathfrak{A}H - W)}\Lambda]$$
$$= \operatorname{Abs}[\theta^{-1} \log |I + \theta P| + 2 \operatorname{Tr}\overline{(\mathfrak{A}H - W)}\Lambda] \tag{17.56}$$

where P is the matrix

$$P(z) = \overline{H(z)} \mathfrak{R} H(z) \tag{17.57}$$

symmetric in that $\bar{P} = P$.

Lemma 17.7.1. *The condition that the Lagrangian form (17.56) be stationary with respect to H can be written*

$$\mathfrak{R}H(I + \theta P)^{-1} + \bar{\mathfrak{A}}\Lambda = \text{neg} \tag{17.58}$$

Proof. We appeal to the fact that, if a square matrix M suffers an increment Δ, then $\log |M|$ suffers the increment $\operatorname{Tr}(M^{-1}\Delta)$. The stationarity condition for $L(H)$ can thus be written

$$\operatorname{Abs} \operatorname{Tr}[\bar{\Delta}\mathfrak{R}H(I + \theta P)^{-1} + \bar{\Delta}\bar{\mathfrak{A}}\Lambda] = 0 \tag{17.59}$$

if Δ is the increment in H. Condition (17.58) then follows from (17.59) by the arguments used for Lemma 17.6.1. □

Relation (17.58) is not linear in H (since P is quadratic) and its solution is not evident. Our aim is to show that (17.58) is satisfied by the LEQG-optimal H determined by (17.45). Henceforth, H will denote the LEQG-optimal value, so that $H(z)$ and $\lambda(z)$ are the solutions of (17.45).

Lemma 17.7.2.

$$\bar{\phi}\phi_0^{-1}\begin{bmatrix} 0 \\ I \end{bmatrix} = \begin{bmatrix} \bar{\chi} \\ I \end{bmatrix} \quad (17.60)$$

where $\bar{\chi} = $ neg, *and the LEQG-optimal H satisfies*

$$\mathfrak{R}H + \bar{\mathfrak{A}}\lambda = \bar{\chi}(W - \theta N\lambda) \quad (17.61)$$

Proof. Relation (17.60) follows directly from the form (17.44) of the canonical factor. Equation (17.45) can be written

$$\phi\begin{bmatrix} H \\ \lambda \end{bmatrix} = \begin{bmatrix} 0 \\ W - \theta N\lambda \end{bmatrix}$$

Pre-multiplying this relation by $\bar{\phi}\phi_0^{-1}$ and appealing to (17.60) we deduce (17.61). □

Lemma 17.7.3. *For the LEQG-optimal H*

$$I + \theta P = \overline{(I - \theta W'\lambda)}(I - \theta W'\lambda_0)^{-1}(I - \theta W'\lambda) \quad (17.62)$$

and this factorization is canonical for $\theta > \theta^*$, *where* θ^* *is the largest value of* θ *for which* $|I + \theta P|$ *has a z-zero on the unit circle.*

Proof. Pre-multiplying equation (17.61) by $\bar{H}$ we obtain

$$P + W'\lambda = \bar{H}\bar{\chi}W(I - \theta W'\lambda)$$

whence

$$I + \theta P = (I + \theta\bar{H}\bar{\chi}W)(I - \theta W'\lambda) \quad (17.63)$$

Now, for $\theta > \theta^*$ the factor $(I - \theta W'\lambda)$ is pole- and zero-free in $|z| \leq 1$, and so is a canonical 'positive' factor. To see this, note that $\lambda(z)$, like $H(z)$, is pole-free in $|z| \leq 1$; so then is $I - \theta W'\lambda$. The determinant $|I - \theta W'\lambda|$ will be zero-free in $|z| \leq 1$ for θ numerically small enough. If θ is decreased from zero then, by (17.63), $|I - \theta W'\lambda|$ cannot develop a zero on the unit circle before $|I + \theta P|$ does. By definition, this occurs first for $\theta = \theta^*$.

By the same argument, the other factor of (17.63) is canonical 'negative'. Since $\bar{P} = P$ and the first factor of (17.63) has absolute term I, factorization (17.63) must imply the symmetric factorization (17.62). □

When the LEQG and entropy criteria are identified then θ^* will, of course, be identified with $\bar{\theta}$, the infimal value of θ at which both optimization problems are well posed.

Theorem 17.7.4. *The LEQG-optimal transfer function H determined by* (17.45) *indeed satisfies the entropy-optimality condition* (17.58), *with Λ determined by*

$$W'\Lambda = -[P(I + \theta P)^{-1}]_+ \qquad (17.64)$$

Proof. The necessity of relation (17.64) follows if we premultiply equation (17.58) by $\bar{H}$. Inserting first expression (17.64) for Λ and then expression (17.61) for $\Re H$ in the left-hand member of (17.58) we deduce that

$$\Re H(I + \theta P)^{-1} + \bar{\mathfrak{A}}\Lambda$$
$$= [\Re H - \bar{\mathfrak{A}}(W')^{-1}P](I + \theta P)^{-1} + \text{neg}$$
$$= [I - \bar{\mathfrak{A}}(W')^{-1}\bar{H}]\Re H(I + \theta P)^{-1} + \text{neg}$$
$$= [I - \bar{\mathfrak{A}}(W')^{-1}\bar{H}][-\bar{\mathfrak{A}}\lambda + \bar{\chi}W(I - \theta W'\lambda)](I + \theta P)^{-1} + \text{neg}$$

Examining this last expression, we see that the first bracket multiplied into $\bar{\mathfrak{A}}$ gives zero, in virtue of (17.39). Substituting for $(I + \theta P)^{-1}$ from (17.62) we are then left with an expression

$$[I - \bar{\mathfrak{A}}(W')^{-1}\bar{H}]\bar{\chi}W(I - \theta W'\lambda_0)(I - \theta\bar{\lambda}W)^{-1} + \text{neg} = \text{neg}$$

Relation (17.58) is thus satisfied by the LEQG-optimal H. □

From (17.62) and (17.64) we deduce that

$$I + \theta W'\Lambda = [(I + \theta P)^{-1}]_+ = [(I - \theta W'\lambda)^{-1}(I - \theta W'\lambda_0)\overline{(I - \theta W'\lambda)}^{-1}]_+$$

By comparing the coefficients of θ on the two sides of this equation we see that the two generating functions of Lagrange multipliers, Λ and λ, are indeed equal in the risk-neutral case, $\theta = 0$. However, they differ otherwise.

17.8 COMPLETION OF THE LEQG/ENTROPY BRIDGE: DEGREES OF OPTIMALITY

Theorem 17.7.4 (which subsumes its risk-neutral version, Theorem 17.6.2) states that the LEQG-optimal transfer function H satisfies the entropy-optimality condition (17.58). H also satisfies the constraints for which the Lagrangian multipliers Λ_j were introduced. However, this does not establish entropy-optimality of H unless condition (17.58) is sufficient as well as necessary for optimality.

Theorem 17.8.1. *Suppose that $\theta > \bar{\theta}$. Then the LEQG-optimal transfer function H determined by* (17.45) *is entropy-minimizing. Suppose also that $W > 0$ and that the system is deviation-sensitive. Then H is the unique entropy-minimizing transfer function.*

Proof. The first condition ensures that $\mathscr{E}(\theta)$, given by (17.40), is convex in H, so that a stationarity condition locates a minimum of $\mathscr{E}(\theta)$. The second condition ensures *strict* convexity within the class of H satisfying the constraint $\mathfrak{A}H = W$. The minimizing value is then unique. □

The condition $W > 0$ seems also to be necessary for the treatment of Section 17.7, in that W^{-1} occurs repeatedly. The condition is certainly a natural one in the H_∞ formulation, if the system is to be proofed against all possible disturbing inputs. However, for the LEQG formulation there is no such requirement. Indeed, the extreme case $W = 0$ (and so $N = 0$) then makes perfect sense, as the problem of deterministic regulation.

Formally, the point is met by the fact that $H(z)$ will have the form $K(z)W$, where $K(z)$ is the transfer function $\varepsilon \to \mathfrak{x}$. Thus, even if W is singular, $K = HW^{-1}$ can still be assigned a finite (although indeterminate) value. This indeterminacy is reflected in the fact that the entropy criterion has the expression

$$\mathscr{E}(\theta) = \frac{1}{2\pi\theta} \int_{-\pi}^{\pi} \log|I + \theta W' \bar{K} \mathfrak{R} K W| \, d\omega$$

in terms of K, and will not be strictly convex in K if W is singular. The extreme case of singularity is when $W = 0$. In this case *any* stabilizing control makes $\mathscr{E}(\theta)$ zero, and so is entropy-minimizing and LEQG-optimal in the average sense.

The reason for this indeterminacy is that average-cost optimality is a so much weaker concept than total-cost optimality. This is the point first made clearly by Blackwell (1962) for Markov decision processes in general. Suppose that the policy which minimizes the total expected cost $\gamma_\pi(\theta)$ has a limit as the horizon h becomes infinite. This policy is then the Blackwell policy, which minimizes not only average expected cost $\delta_\pi(\theta)$ but also the transient cost incurred in passing from arbitrary initial conditions to a stationary regime. The policy associated with the LEQG-optimal H determined by (17.45) has this character. An average-optimal policy will not necessarily have this character unless all modes of variation continue to be stimulated under it. For our problem, this requires that $W > 0$. In other cases there may be many policies which are average-optimal but do not minimize transient costs.

17.9 NOTES ON THE LITERATURE

For references on H_∞ theory see e.g. Zames (1981) and Francis (1987). For work on entropy minimization, see e.g. Arov and Krein (1983), and Dym and Gohberg (1986). As stated, the Glover–Doyle theorem, Theorem 17.3.1, is to be found in Glover and Doyle (1988). The material of Section 17.4 is original, although doubtless implicit in the literature. The material of Sections 17.5–8 was presented in a more abbreviated form in Whittle (1989b).

CHAPTER 18

Variants

18.1 DISCOUNTING

We have hitherto not considered discounted cost criteria, but have minimized either the total cost over a finite horizon or the average cost under stationary rules. However, suppose we consider the discounted cost function

$$\mathbb{C} = \sum_t \beta^t c(x_t, u_t) \tag{18.1}$$

with c having the familiar homogeneous quadratic form (3.9). Here β is the discount factor, usually chosen in the range $0 \leq \beta \leq 1$, the choice $\beta < 1$ corresponding to *strict* discounting. We assume a plant equation of general order

$$\mathscr{A}(\mathscr{T})x + \mathscr{B}(\mathscr{T})u = 0 \tag{18.2}$$

For the moment, we assume this deterministic and observation perfect. If we associate a discounted Lagrangian multiplier $\beta^t \lambda_t$ with the plant equation (18.2) at t then minimization of the Lagrangian form

$$\sum_\tau \beta^\tau [c(x, u) + \lambda'(\mathscr{A}x + \mathscr{B}u)]_\tau \tag{18.3}$$

leads to the equation system

$$\Phi(\mathscr{T}) \begin{bmatrix} x \\ u \\ \lambda \end{bmatrix}_\tau^{(t)} = 0 \qquad (\tau \geq t) \tag{18.4}$$

with Φ now having the asymmetric form

$$\Phi(z) = \begin{bmatrix} R & S' & \mathscr{A}(\beta z^{-1})' \\ S & Q & \mathscr{B}(\beta z^{-1})' \\ \mathscr{A}(z) & \mathscr{B}(z) & 0 \end{bmatrix} \tag{18.5}$$

This deviates from the previous pattern (c.f. (10.13)) in that the discount factor β modifies the entries which were previously $\overline{\mathscr{A}}$ and $\overline{\mathscr{B}}$. The equation system (18.4) can still be reduced to an explicit solution for the optimal infinite-horizon control by canonical factorization of Φ, although the factorization will now be an asymmetric one.

Suppose we now add process noise and imperfect observation, and assume ourselves in the risk-neutral case. Then we know from the certainty-equivalence principle that process noise, if it is white, will not affect the open-loop form of the optimal control rule. Furthermore, the effect of imperfect observation is just that one substitutes the ML estimates $x_\tau^{(t)}$ for x_τ in the control rule ($\tau \leq t$). Discounting will not affect the calculation of the estimates $x_\tau^{(t)}$ or their covariances.

However, in the risk-sensitive case the effect of discounting is radical, because the component $\mathbb{C}$ of stress has a discounted form, while the component $\mathbb{D}$ does not. This means that the optimal policy must be non-stationary, because, as one advances into the future, the minimization of $\mathbb{C}$ becomes increasingly subject to the prior minimization of $\mathbb{D}$. Thus, the effect is very much as if θ would decrease in value as time progresses—i.e as though one approached the risk-neutral case, and at an exponential rate.

For example, consider a discounted version of the scalar regulation problem considered in Section 9.1. If the value function from time t is $\Pi_t x_t^2 = \beta^t P_t x_t^2$ then the optimal control is

$$u_t = \frac{ABx_t}{Q\beta^t(\Pi_{t+1}^{-1} + B^2\theta N/(Q\beta^t))} = \frac{\beta ABP_{t+1}x_t}{Q + \beta P_{t+1}(B^2 + \theta NQ\beta^t)} \qquad (18.6)$$

where P_t obeys the recursion

$$P_t = R + \frac{\beta A^2 Q P_{t+1}}{Q + \beta P_{t+1}(B^2 + \theta NQ\beta^t)} \qquad (18.7)$$

Thus θ occurs as $\theta\beta^t$ and, as t becomes large, relations (18.6) and (18.7) both converge to the risk-neutral form from an initially risk-sensitive form.

Equations (18.4) and (18.5) seem to be stated first in Whittle (1983, p. 72), although relations very like them are found in Telser and Graves (1972). Bouakiz and Sobel (1984) appear to have been the first to observe that an optimal risk-sensitive policy is necessarily non-stationary in the discounted case.

18.2 RELAXATION OF THE LEQG ASSUMPTIONS ON THE STATE VARIABLE

Consider a state-structured process in continuous time. It was shown by Whittle and Gait (1970) that the LQG assumptions for such a process could be considerably weakened as far as the state variable was concerned, and interesting results still obtained. Kuhn (1985) generalized these conclusions to the LEQG case; we shall outline his analysis. See also Section 18.5.

Let us suppose that the process has a currently observed vector state variable x which obeys the first-order stochastic differential equation

$$\dot{x} = A(x, t) + Bu + \varepsilon \qquad (18.8)$$

Here ε is, as ever, vector white noise with zero mean and power matrix N. We

18.2 RELAXATION OF THE LEQG ASSUMPTIONS ON THE STATE VARIABLE

assume that the process terminates at the time T when the compound variable $(x(t), t)$ first enters a prescribed stopping set D, and that, if the process has not stopped by time t, then the cost incurred from t is

$$\mathbb{C}(t) = \int_t^T [\tfrac{1}{2} u'Qu + m(x, \tau)] \, d\tau + \mathbb{K}(x(T), T) \tag{18.9}$$

At time t the stopping time T is, of course, like the future path of the process, a random variable.

We see from (18.8) and (18.9) that the only linear/quadratic hypotheses we have made are that the plant equation should be linear in u and the instantaneous cost quadratic in u. The terms $A(x, t)$ and $m(x, t)$ allow a general dependence of dynamics and costs upon (x, t), and the stopping rule is similarly general.

We shall adopt a risk-sensitive criterion, and in fact define a value function

$$F(x, t) = \inf_\pi \left[-\theta^{-1} \log E_\pi(\exp(-\theta \mathbb{C}(t)) | x(t) = x) \right] \tag{18.10}$$

Let us, for notational simplicity, define

$$f(x, u, t) = A(x, t) + Bu$$

$$g(x, u, t) = \tfrac{1}{2} u'Qu + m(x, t)$$

and use a subscript notation for partial differentials. Thus, F_t denotes $\partial F(x, t)/\partial t$, and F_x and F_{xx} are the corresponding row vector and matrix of first- and second-order x-derivatives, respectively.

Lemma 18.2.1. *Under the hypotheses above the value function F obeys the optimality equation*

$$\inf_u \left[g + F_t + F_x f - \tfrac{1}{2} F_x(\theta N) F_x' + \tfrac{1}{2} \mathrm{Tr}(N F_{xx}) \right] = 0 \tag{18.11}$$

outside D, with $F = \mathbb{K}$ in D. The optimal control is thus given by

$$u = -Q^{-1} B' F_x' \tag{18.12}$$

The proof is routine; we refer to Kuhn's paper.

Let us now make the assumption that there exists a scalar α such that

$$\alpha N = BQ^{-1}B' + \theta N \tag{18.13}$$

That is, the noise power matrix N is a scalar multiple of the effective control power matrix J.

Theorem 18.2.2. *Under the additional assumption (18.13) one can assert that*

$$F(x, t) = -\alpha^{-1} \log G(x, t)$$

where

$$G(x,t) = E\left[\exp\left(-\alpha \int_t^T m(x,\tau)\,d\tau - \alpha \mathbb{K}(x(T),T)\right)\bigg| x(t) = x\right] \quad (18.14)$$

and the x-path and termination time T in (18.14) are those for the uncontrolled process governed by

$$\dot{x} = A(x,t) + \varepsilon \quad (18.15)$$

Proof. If we perform the u-minimization in (18.11) we obtain the equation

$$m + F_t + F_x A - \tfrac{1}{2} F_x J F_x' + \tfrac{1}{2}\text{Tr}(N F_{xx}) = 0 \quad (18.16)$$

The transformation to G linearizes this equation to

$$-\alpha m G + G_t + G_x A + \tfrac{1}{2}\text{tr}(N G_{xx}) = 0 \quad (18.17)$$

This holds outside D, with $G = e^{-\alpha \mathbb{K}}$ within D. However, this is exactly the equation and boundary condition satisfied by G as defined in (18.14). $\square$

The optimal control problem has thus been reduced, under hypothesis (18.13), to the calculation of first-passage statistics to D for an uncontrolled problem. Kuhn analyses the 'scalar landing problem' assuming risk-sensitivity: the calculation of the ideal landing path for an inertialess aircraft. We give another example in Exercise 18.2.1.

If (18.13) holds then $BQ^{-1}B'$ is a scalar multiple of N, say $\alpha_0 N$. If m and $\mathbb{K}$ are non-negative then expectation (18.14) will, in general, diverge for α less than some critical value, say α_1. We see then that the critical ('breakdown') value of θ is $\alpha_1 - \alpha_0$. For lower values of θ expected costs are infinite under all policies.

18.2.1 Exercise and comments

(1) *'Survival'*. This is the problem considered by Lefebvre and Whittle (1988). The state variable x is scalar, satisfying simply $\dot{x} = Bu + \varepsilon$. The initial value x lies in the interval $(-a, a)$ and the aim is to so control motion that

$$\mathbb{C} = -T + \frac{1}{2}\int^T Qu^2\,d\tau$$

is small, where T is the moment when x first reaches one of the boundary values $\pm a$. One is then trying to maximize survival time subject to a constraint on control effort. One could imagine the problem as representing a low-flying aircraft (again inertialess, unfortunately!) continuing its flight as long as possible before it encounters either the ground or the ceiling at which it will be detected by radar.

In the notation of the section we have $A = 0$ and $m = -1$. The problem is time invariant, so $G_t = 0$ and equation (18.17) becomes

$$\alpha G + \tfrac{1}{2} N G_{xx} = 0 \qquad (18.18)$$

where

$$\alpha N = B^2 Q^{-1} + \theta N$$

Equation (18.18) is to be solved in $(-a, a)$ subject to $G(\pm a) = 1$. The solution is

$$G(x) = \frac{\cos(\omega x)}{\cos(\omega a)} \qquad (18.19)$$

where

$$\omega = \sqrt{(2\alpha/N)} = \sqrt{[2(J_0 N^{-2} + \theta N^{-1})]}$$

The optimal control is

$$u = -Q^{-1} B F_x = \frac{B G_x}{Q \alpha G} = -\frac{B \omega}{Q \alpha} \tan(\omega x)$$

For validity of solution (18.19) we require that $\cos(\omega x)$ should not change sign in $(-a, a)$, so that $\omega a < \pi/2$, or

$$\theta < \frac{\pi^2 N}{8a^2} - \frac{B^2}{NQ}$$

Therefore, as in some of the examples of Section 7.8, we find that θ is intrinsically bounded from above rather than from below. Roughly speaking, $E(e^{\theta T})$ will be infinite unless θ is smaller than the rate at which survival probability decays with time. For θ exceeding this value the optimizer is euphoric in that the 'expected reward' is infinite.

The familiar lower bound on θ implied by $J_0 + \theta N > 0$ has a different consequence for this problem. If $J_0 + \theta N$ becomes negative then so does α, so that ω becomes imaginary and the cos terms in solution (18.19) are replaced by cosh functions. This is formally acceptable. A negative value of J indeed implies ineffective control and early termination, but with a well-defined value function.

18.3 AGENTS WITH DIFFERING RISK SENSITIVITIES

Suppose the control u can be decomposed into sub-controls u_i ($i = 1, 2, \ldots, s$) in terms of which the plant equation becomes

$$x_t = A x_{t-1} + \sum_i B_i u_{i,t-1} + \varepsilon_t \qquad (18.20)$$

and the instantaneous cost function

$$c(x, u) = x' R x + \sum_i u_i' Q u_i \qquad (18.21)$$

We regard u_i as being the control exerted by the ith of s agents, who will choose this optimally on the basis of cost function (18.21). We assume that all agents have the same information, including that on past controls.

Under these suppositions the agents are acting as a team, and the joint optimization reduces to an optimization as by a single agent with respect to the total control variable u.

Suppose, however, that in a risk-sensitive formulation we allow the risk-sensitivity parameter to vary between agents, agent i having parameter θ_i. Then, although the agents have common information and a common cost function $\mathbb{C}$, they optimize according to different criteria, $-2\theta_i^{-1}\log(E\exp[-(\theta_i/2)\mathbb{C}])$. The agents will then have differing value functions. However, the joint optimum is still achieved if each agent optimizes control by appealing to the optimality equation as he sees it, and the RSCEP will still imply that the value function has a quadratic form satisfying the analogue of equation (7.18).

Therefore let us assume that all agents observe the current state. We can then expect that the x-dependent part of the infinite-horizon value function for agent i has the form $x'\Pi_i x$, and have the set of Riccati equations implied by

$$x'\Pi_i x = \min_{u_i}\text{ext}_{\varepsilon}\left[c(x,u) + \varepsilon'(\theta_i N)^{-1}\varepsilon + \left(Ax + \sum_k B_k u_k + \varepsilon\right)'\Pi_i\left(Ax + \sum_k B_k u_k + \varepsilon\right)\right] \quad (i = 1, 2, \ldots, s) \tag{18.22}$$

The subsequent treatment simplifies greatly if we consider rather the continuous-time version, in which (18.20) is replaced by

$$\dot{x} = Ax + \sum_i B_i u_i + \varepsilon \tag{18.23}$$

and (18.22) by

$$\min_{u_i}\text{ext}_{\varepsilon}\left[c(x,u) + \varepsilon'(\theta_i N)^{-1}\varepsilon + 2x'\Pi_i\left(Ax + \sum_k B_k u_k + \varepsilon\right)\right] = 0 \quad (i = 1, 2, \ldots, s) \tag{18.24}$$

Equation (18.24) implies the evaluations

$$u_i = -Q^{-1}B_i'\Pi_i x, \qquad \varepsilon = -\theta_i N\Pi_i x \tag{18.25}$$

and the system of Riccati equations

$$R + A'\Pi_i + \Pi_i A - \theta_i\Pi_i N\Pi_i + \sum_k [\Pi_k J_k \Pi_k - \Pi_k J_k \Pi_i - \Pi_i J_k \Pi_k] = 0$$

$$(i = 1, 2, \ldots, s) \tag{18.26}$$

Here

$$J_k = B_k Q_k^{-1} B_k' \tag{18.27}$$

18.3 AGENTS WITH DIFFERING RISK SENSITIVITIES

In general, there is no recourse but to solve the system of coupled Riccati equations (18.26). However, a first-order perturbation approximation to these equations has a strikingly simple form. Suppose that the θ_i vary only slightly about a value θ, so that

$$\theta_i = \theta + \delta_i \tag{18.28}$$

where the δ_i are small. Set correspondingly

$$\Pi_i = \Pi + \Delta_i \tag{18.29}$$

where Π solves the Riccati equation for given θ. That is,

$$R + A'\Pi + \Pi A - \Pi J \Pi = 0 \tag{18.30}$$

where

$$J = \theta N + \sum_k J_k \tag{18.31}$$

Substituting (18.28) and (18.29) into (18.26) and retaining only first-order terms in δ and Δ, we obtain the equations

$$\Gamma' \Delta_i + \Delta_i \Gamma = \delta_i \Pi N \Pi \tag{18.32}$$

where

$$\Gamma = A - J\Pi \tag{18.33}$$

is the gain matrix for the optimized system with $\theta_i \equiv \theta$. We can then state

Theorem 18.3.1. *If terms of higher degree than the first in $\delta_i = \theta_i - \theta$ and $\Delta_i = \Pi_i - \Pi$ are neglected in the system of Riccati equations (18.26) then these equations have the solution*

$$\Pi_i = \Pi + \delta_i D \tag{18.34}$$

where D is the symmetric solution of

$$\Gamma' D + D \Gamma = \Pi N \Pi \tag{18.35}$$

The matrix D can be identified as $\partial \Pi / \partial \theta$, where Π is the root of (18.30). The interest of the theorem is then that, to first order, Π_i is affected by the variation of θ to θ_i just as Π would be, but not by variation in the θ-values for other agents. Since Γ is a continuous-time stability matrix and $\Pi N \Pi$ is non-negative definite, then D is non-positive definite, reflecting the fact that Π is non-increasing in θ. By use of (18.25) and the perturbed value of Π_i we obtain the perturbed value of the optimal control.

In discrete time we can expect that Theorem 18.3.1 will still hold with D now determined by

$$D - \Gamma' D \Gamma + \Pi N \Pi = 0$$

Γ and Π being the evaluations for the problem with constant θ. The modification in control is a little more difficult to calculate, because the controls for different agents at the same time interact; see Exercise 18.3.1.

18.3.1 Exercise and comments

(1) Show that in the discrete-time case the perturbation to the optimal action for agent i due to variation of θ_i is

$$\delta u_i = -Q_i^{-1} B_i' \left[\delta_i - (\Pi^{-1} + J)^{-1} \sum_k J_k \delta_k \right] (\Pi^{-1} + \theta N)^{-1} (N - \Pi^{-1} D \Pi^{-1})$$
$$\times (\Pi^{-1} + J)^{-1} A x$$

18.4 NOTES ON THE LITERATURE

There are, of course, a great many papers dealing with risk sensitivity in some sense, particularly in the economic literature. The ones that we would regard as clearly relevant to this work are those that, to some degree, take the LEQG approach, i.e. consider the exponential-of quadratic criterion. We quote the few of which we are aware (in addition to papers already quoted); there are doubtless others that are relevant and valuable.

A series of papers by Karp (see the References) pays particular attention to the case of many agents. He discusses dynamic hedging, risk aversion in a dynamic trading game (very close to the topic of the previous section) and the endogenous stability of economic systems. Papers by Caravani (1986) and Caravani and Papavassilopoulos (1987) consider risk sensitivity in both economic and game contexts. Van der Ploeg (1983, 1984) provided an early economic application and development of the LEQG solutions derived in Whittle (1981). In particular, he derived an unemployment/inflation trade-off and considered the stabilization of an open economy multiplier-accelerator model, in both cases in a risk-sensitive formulation.

18.5 THE PATH-INTEGRAL FORMULATION AND THE STOCHASTIC MAXIMUM PRINCIPLE FOR NON-LQG PROCESSES

We have worked totally with processes having the LQG property (with only the minor variations of Section 18.2) and our results have been exact. One wonders if there is any general extension to non-LQG processes. It turns out that there is. To achieve this extension one must appeal to two features: (1) risk-sensitivity, in that the criterion must be the expectation of an exponential function of costs, and (2) results in the non-LQG case are to be considered as large-deviation approximations. This work is new, and is set out in more detail in Whittle (1990).

18.5 THE PATH-INTEGRAL FORMULATION

Suppose that $x(t)$ is the vector state variable of a Markov decision process in continuous time, and that this is currently observable. Let Δx be the increment $x(t + \Delta t) - x(t)$ in time Δt. Define the *derivate characteristic function* (d.c.f.)

$$H(x, u, \alpha) = \lim_{\Delta t \downarrow 0} \frac{E(\exp[\alpha' \Delta x] - 1 | x(t) = x, u(t) = u)}{\Delta t}$$

a concept due to Bartlett. Suppose now that the process is modified to be almost deterministic, in that its d.c.f is scaled to $VH(x, u, \alpha/V)$, where V is a large scale parameter.

Suppose that the criterion from time t is $-\theta^{-1} \log(E_\pi[-\theta \mathbb{C}_t])$, where

$$\mathbb{C}_t = \int_t c(x, u) \, d\tau + \text{(terminal terms)}$$

is the cost function from time t. Then application of large-deviation theory shows that

$$\inf_\pi [-\theta^{-1} \log\{E_\pi(\exp[-\theta \mathbb{C}_t]) | x(t) = x)\}] = VF(x, t) + 0(V)$$

where $F(x, t)$, the scaled asymptotic value function, is the extremum of the path integral

$$\mathbb{I} = \int_t [c(x, u) + \lambda' \dot{x} - \theta^{-1} H(x, u, \theta \lambda)] \, d\tau + \text{(terminal terms)}. \quad (18.36)$$

The extremum is with respect to the (x, u, λ)-path, subject only to $x(t) = x$, the nature of the extremum being that asserted on p. 135. The value of $u(t)$ thus determined is optimal.

In (18.36) we recognise a generalisation of the path-integral (15.1) to the non-LQG case (with the difference of a factor of 2, due to slightly changed conventions). The penalty for the generalisation is that evaluations are asymptotic for large V, rather than exact.

The stationarity conditions for the integral (18.36) with respect to (x, u, λ) constitute the stochastic maximum principle and determine a minimal-stress future path. The scaled value function $F(x, t)$ obeys the dynamic programming equation

$$\inf_u \left[c(x, u) + \frac{\partial F}{\partial t} - \theta^{-1} H\left(x, u, -\theta \frac{\partial F}{\partial x}\right) \right] = 0.$$

The form of the generalisation of these results to the non-Markov case and to the case of imperfect observation will be apparent from the text.

APPENDIX 1

Abbreviations

We list a number of abbreviations taken as standard throughout the text.

- CEP   Certainty-equivalence principle (Section 2.5)
- LEQG   Linear/exponential-of-quadratic/Gaussian (Sections 1.2 and 6.1)
- LQG   Linear/quadratic/Gaussian (Sections 1.1 and 2.2)
- LSE   Least-square estimate, to be identified also with linear least estimate (Section 2.4)
- MLE   Maximum likelihood estimate (Section 2.4)
- MSE   Minimum stress estimate, although more properly characterized as extremal stress estimate, with the mode of extremum appropriately defined (Section 6.3)
- RSCEP Risk-sensitive certainty-equivalence principle (Section 6.2).

APPENDIX 2

Notation and a list of symbols

In discrete-time models the time variable t is assumed to take signed integer values; in continuous-time models it may take any real value.

In discrete time the value of a variable x at time t is denoted x_t. More generally, if $(\ldots)$ is a bracketed expression then $(\ldots)_t$ denotes this expression with all quantities in the bracket evaluated at time t. The expression $x_\tau^{(t)}$ denotes the best (minimal stress) estimate of x_τ based on information available at time t (Sections 2.4 and 6.3).

If $\{x_t\}$ is a sequence of variables in discrete time then X_t denotes the history of this sequence up to time t, so that $X_t = \{x_\tau; \tau \leq t\}$. The starting point of this sequence may depend upon circumstances.

A matrix is denoted by an italic capital: A. Its transpose is denoted by A'. If Q is a matrix then the relations $Q > 0$ and $Q \geq 0$ indicate, respectively, that Q is positive-definite and positive-semi-definite (and so understood to be symmetric).

Operators are generally denoted by a script capital (Section 1.3). Important special operators are the time shift operator $\mathscr{T}$ with effect $\mathscr{T} x_t = x_{t-1}$ and the time differential operator $\mathscr{D}$ with effect $\mathscr{D} x = dx/dt$ (also denoted $\dot{x}$). A symbol $\mathscr{A}$ will denote a distributed lag operator $\mathscr{A}(\mathscr{T}) = \sum_j A_j \mathscr{T}^j$ in discrete time or a differential operator $\mathscr{A}(\mathscr{D}) = \sum_j A_j \mathscr{D}^j$ in continuous time. The *conjugate* $\bar{\mathscr{A}}$ of $\mathscr{A}$ is defined as $\mathscr{A}(\mathscr{T}^{-1})' = \sum_j A_j' \mathscr{T}^{-j}$ and as $\mathscr{A}(-\mathscr{D})' = \sum_j \mathscr{A}_j'(-\mathscr{D})^j$ in these two respective cases. We shall often consider the corresponding generating function $\mathscr{A}(z) = \sum_j A_j z^j$, a function of the complex scalar argument z. This will often also be denoted $\mathscr{A}$, the interpretation (operator or generating function) being clear from the context. In this case the conjugate $\bar{\mathscr{A}}$ is correspondingly defined as $\mathscr{A}(z^{-1})'$ or $\mathscr{A}(-z)'$ in the discrete- and continuous-time cases, respectively.

The shell or Gill characters $\mathbb{S}$, $\mathbb{C}$ and $\mathbb{D}$ are used to denote stress and the cost and discrepancy components of stress (Section 6.2). In particular, $\mathbb{C}_h$ and $\mathbb{D}_0$ are used to denote terminal cost and initial discrepancy, $-2(\text{log-likelihood})$, respectively. A related notation is the use of $\mathbb{I}$ to denote a path integral (Sections 10.2 and 10.3).

Gothic or fraktur symbols $\mathfrak{A}$, $\mathfrak{R}$ and $\mathfrak{x}$ are used for the system notation defined in Section 10.7, and used in Chapters 13 and 17, in particular.

The symbols $+$ and $-$ are used as subscripts to denote canonical factors of appropriate characters, as in $\Phi = \Phi_- \Phi_0 \Phi_+$ (Sections 3.7, 5.4 and 10.4). The notation $[f(z)]_+$ denotes the operation of truncating the power series representation of $f(z)$ on the unit circle to include only non-negative powers (Section 17.1).

The notations max or min before an expression indicate the operation of taking a maximum or minimum: e.g. $\min_u$ denotes the operation of minimizing with respect to the variable u and correspondingly for sup and inf. The notation stat indicates the operation of evaluating the expression at a stationary point (with respect to variables indicated) and ext denotes the evaluation at an extremum, where the nature of the extremum is specified in the context (Section 6.2).

The expectation operator is denoted by E, and E_π denotes the expectation under a policy π. Probability measure is indicated by $P(\cdot)$ and probability density sometimes by $f(\cdot)$. Conditional versions of these quantities are denoted by $E(\cdot|\cdot)$, $P(\cdot|\cdot)$ and $f(\cdot|\cdot)$, respectively.

The covariance matrix of a vector random variable ε is written $\mathrm{cov}(\varepsilon)$. The covariance matrix between two such variables, ε and η is written $\mathrm{cov}(\varepsilon, \eta)$, so that $\mathrm{cov}(\varepsilon, \varepsilon) = \mathrm{cov}(\varepsilon)$.

A full range of symbols is used, some of them perforce in more than one sense. We therefore make a systematizing list.

LIST OF SYMBOLS (WITH SECTION REFERENCES)

A, B, C	Coefficient matrices in the plant and observation equations (1.1)
$\mathcal{A}, \mathcal{B}, \mathcal{C}$	Operator versions of these for higher-order schemes (1.3)
$\mathfrak{A}$	System version of A (10.7)
$\bar{\mathcal{A}}$, etc.	Conjugate of $\mathcal{A}$, etc (10.2, 15.3, 16.2)
$\mathbb{C}$	Cost function (1.1, 2.2, 2.6)
$c(x, u)$	Instantaneous cost function (1.1, 3.2)
D	Component of canonical factor (13.1, 13.2)
	The matrix $\partial \Pi / \partial \theta$ (18.3)
$\mathcal{D}$	Time differential operator (1.3, 15.1)
$\mathbb{D}$	The discrepancy component of stress (2.6, 6.2)
d	The instantaneous discrepancy (7.1)
E	Expectation operator (1.1)
E_π	Expectation under policy π (1.1)
$\mathcal{E}$	The projection operator of least square approximation (2.4)
F	Value function (2.3, 3.1)
	Future stress (7.2)
f	The operator for the backward Riccati equation (3.2)
$\tilde{f}$	The operator defined in equation (7.21)
f_Π	The linear operator defined by $f\Pi$ (9.4)
G	A value function (2.3)
	Coefficient of the quadratic terms in a path integral (10.5)
	A transfer function (17.2)
H	The innovation coefficient in the Kalman filter (4.2)

LIST OF SYMBOLS (WITH SECTION REFERENCES)

	The constant factor in a canonical factorization (12.1)
	The matrix coefficient of highest-order derivatives (15.3)
	The system transfer function (17.2)
h	Horizon (2.1)
I	The identity matrix
$\mathbb{I}$	Path integral (3.6, 10.2, 10.3, 10.5, 15.2)
J	The control-power matrix (3.4, 7.4)
K	The matrix coefficient in the optimal feedback control (3.2)
$\mathbb{K}$	A terminal cost function (18.2)
L	Covariance matrix of plant and observation noise (4.1, 4.5)
$\mathscr{L}$	The one-step operator occurring in the optimality equation (3.1)
l	A variable conjugate to x (5.2)
M	Covariance matrix of observation noise (4.1, 4.5)
m	A variable conjugate to u (5.2)
	Dimension of the control variable u (1.1)
N	Covariance matrix of plant noise (1.1, 4.1)
n	Dimension of the process variable x (1.1)
P	Past stress (7.2)
	Probability measure
p	Order of dynamics (10.1, 12.1)
	The operator for the forward Riccati equation (4.2)
Q	Matrix of control costs (1.1)
R	Matrix of process-variable costs (1.1)
$\mathfrak{R}$	The system version of R (10.7)
r	The dimension of the observation y (1.1)
S	Cross-matrix for control and process-variable costs (1.1)
$\mathbb{S}$	Stress (6.2)
s	Time-to-go (3.1)
$\mathscr{T}$	The backward shift operator (1.3)
T	Stopping time (18.2)
t	Time
U	Control history (1.1, 2.1)
u	Control (action, decision) variable (1.1, 1.3)
$\bar{u}$	Reference value of u (3.3)
V	Covariance matrix or a risk-sensitive analogue (4.2, 8.1)
$v(\lambda, \mu)$	The information analogue of $c(x, u)$ (5.2, 10.2)
W	Information history (1.1, 2.2, 2.3)
	A noise-scaling matrix (17.2)
X	Process history (1.1, 2.1)
x	Process variable; state variable if one exists (1.1, 1.3)
$\bar{x}$	Reference value of x (3.3)
$\hat{x}$	Estimate of x based on past stress (4.2)
$\check{x}$	Estimate of x based on total stress (7.2)

NOTATION AND A LIST OF SYMBOLS

x	System variable (10.7)
y	Observation (1.1, 1.3)
$\tilde{y}$	Reduced observation (2.1)
z	Argument of generating function
α	Deterministic disturbance to the plant equation (3.3)
Γ	Gain matrix (2.3, 7.4)
$\tilde{\Gamma}$	Predictive gain matrix (7.4)
γ	The LEQG criterion function (1.2, 6.1)
	The constant term in a value function (3.2, 3.3, 7.5)
Δ	Estimation error (1.1, 4.2)
	An increment (3.8, 17.2)
δ	The LEQG average criterion (17.3)
ε	Plant noise (1.1, 1.3)
ζ	The variable of a forward path integral (3.7, 10.4, 10.5)
η	Observation noise (1.1, 1.3)
θ	Risk-sensitivity parameter (1.2, 6.1)
$\bar{\theta}$	A value of θ at which the problem becomes ill posed (6.4)
λ	The Lagrange multiplier for the plant constraint (3.7, 10.2, 10.3)
μ	The Lagrange multiplier for the observation constraint (10.3)
ν	A standardized plant noise variable (17.2)
ξ	Vector of all noise inputs (2.2)
Π	Matrix of a value function (3.2)
π	A policy (1.1, 2.3)
ρ	Coefficient of the linear term in the backward path integral (5.3, 10.4)
σ	Coefficient of the linear term in the value function (3.3)
	A matrix element and its generating function (17.3)
τ	A running time variable
Φ	A matrix of operators (or generating functions) occurring in forward optimization (3.7, 10.2, 10.3)
$\Phi(w, z)$	The generating function defined in equation (12.21)
ϕ	A canonical factor of Φ (12.1)
Ψ	A matrix of operators (or generating functions) occurring in backward optimization (5.3, 10.3)
ψ	A canonical factor of Ψ (10.4)
χ	The operator $\tilde{ff}$ of the risk-sensitive Riccati equation (9.1)
	The variable of a backward path integral (10.4)
Ω	Information gain matrix (4.2)
ω	Coefficient of the linear term in the forward path integral (10.4, 10.5)
	Frequency (1.4, 17.3)

APPENDIX 3

Optimal estimation and the Gauss–Markov theorem

As in Section 2.4, assume x and y jointly random vectors of zero mean, the unobservable x to be estimated in terms of the observable y. If these are normally distributed then they have joint density

$$f(x, y) \propto \exp(-\tfrac{1}{2}\mathbb{D}) \tag{A3.1}$$

where

$$\mathbb{D} = \begin{bmatrix} x \\ y \end{bmatrix}' \begin{bmatrix} V_{xx} & V_{yy} \\ V_{yx} & V_{yy} \end{bmatrix}^{-1} \begin{bmatrix} x \\ y \end{bmatrix} = \begin{bmatrix} x \\ y \end{bmatrix}' \begin{bmatrix} I_{xx} & I_{xy} \\ I_{yx} & I_{yy} \end{bmatrix} \begin{bmatrix} x \\ y \end{bmatrix} \tag{A3.2}$$

Here I_{xx}, etc. denote information matrices, not identity matrices.

We now prove the Gauss-Markov theorem, which establishes the equivalence of three characterizations of the estimate $\hat{x}$: the ML and LS characterizations and the orthogonality condition

$$\hat{x} - x \perp y \tag{A3.3}$$

Proof of Theorem 2.4.1 (the Gauss–Markov theorem). The MLE $\hat{x}$ is determined by $\partial \mathbb{D}/\partial x = 0$ or

$$I_{xx}\hat{x} + I_{xy}y = 0$$

Regarding $\mathbb{D}$ as a quadratic form in x and completing the square we then have

$$\mathbb{D} = (x - \hat{x})' I_{xx} (x - \hat{x}) + d(y) \tag{A3.4}$$

where $d(y)$ is a function of y alone. Decomposition (A3.4) implies that $x - \hat{x}$ and y are statistically independent, whence the orthogonality relation (A3.3) and the characterization $\hat{x} = E(x|y)$ follow.

Let $\bar{x} = by$ be any other linear estimate of x in terms of y. It follows from (A3.3) that $\operatorname{cov}(\hat{x} - x, \bar{x} - x)$ is independent of b, so that, in particular,

$$\operatorname{cov}(\hat{x} - x, \bar{x} - x) = \operatorname{cov}(\hat{x} - x, \hat{x} - x) = \operatorname{cov}(\Delta)$$

where $\Delta = \hat{x} - x$. Setting $\delta = \bar{x} - x$ we then have

$$0 \le \operatorname{cov}(\delta - \Delta) = \operatorname{cov}(\delta) - \operatorname{cov}(\delta, \Delta) - \operatorname{cov}(\Delta, \delta) + \operatorname{cov}(\Delta)$$

$$= \operatorname{cov}(\delta) - \operatorname{cov}(\Delta)$$

OPTIMAL ESTIMATION AND THE GAUSS–MARKOV THEOREM

This conclusion, that

$$\text{cov}(\delta) \geq \text{cov}(\Delta)$$

establishes that $\hat{x}$ is also a LSE, because it implies that $E(\delta' M \delta) \geq E(\Delta' M \Delta)$ for any $M > 0$. Furthermore, this conclusion followed entirely from the orthogonality property (A3.3), which is thus sufficient for the LS property.

We have thus established that the ML property implies (A3.3) which implies the LS property. The reverse sequence of implications is immediate. If $\hat{x} = by$ is a LSE of x then stationarity of $E|x - by|^2$ with respect to b implies (A3.3). This latter relation implies that $x - \hat{x}$ and y are independent (in the Gaussian case) which implies the factorization (A3.4), this then implying that $\hat{x}$ is a MLE. □

APPENDIX 4

The Hamiltonian formalism

We should justify the use of the term 'Hamiltonian' for the path-integral formalism of Chapter 10. The classic use of the term derives from Newtonian dynamics, which characterizes the state of the system by two vector variables, x and λ, whose variation in time is such as to render the path integral

$$\mathbb{I} = \int [H(x, \lambda) + \lambda' \dot{x}] dt \qquad (A4.1)$$

extreme. Here H is the Hamiltonian of the system. The variables x and λ would usually be denoted by q and p and interpreted as a position vector and its conjugate momentum, respectively. We use the x, λ notation to exhibit the parallel with the control-optimization problem.

The stationarity equations for the path integral will be

$$\dot{x} = -\frac{\partial H}{\partial \lambda} \qquad \dot{\lambda} = \frac{\partial H}{\partial x} \qquad (A4.2)$$

If the Hamiltonian has the quadratic form

$$H(x, \lambda) = \begin{bmatrix} x \\ \lambda \end{bmatrix}' \begin{bmatrix} H_{11} & H_{12} \\ H_{21} & H_{22} \end{bmatrix} \begin{bmatrix} x \\ \lambda \end{bmatrix} \qquad (A4.3)$$

then the Hamiltonian stationarity conditions (A4.2) take the linear form

$$\Phi(\mathscr{D}) \begin{bmatrix} x \\ \lambda \end{bmatrix} = 0 \qquad (A4.4)$$

where

$$\Phi(\mathscr{D}) = \begin{bmatrix} H_{11} & H_{12} - \mathscr{D} \\ H_{21} + \mathscr{D} & H_{22} \end{bmatrix} \qquad (A4.5)$$

The parallel with the continuous-time formalism of Chapter 15 will now be clear; the Hamiltonian character of the system is reflected in the self-conjugacy, $\bar{\Phi} = \Phi$, of matrix (A4.5).

Explicitly, the control problem of minimizing $\int c(x, u) dt$ subject to a plant equation

$$\dot{x} = a(x, u) \qquad (A4.6)$$

can be formulated as the extremization of the path integral

$$\mathbb{I} = \int \left[c(x,u) + \lambda'(\dot{x} - a(x,u)) \right] dt \tag{A4.7}$$

which is of the form (A4.1) with Hamiltonian

$$H(x,u,\lambda) = c(x,u) - \lambda' a(x,u) \tag{A4.8}$$

There is also an extra variable u, which does not disturb Hamiltonian structure, because the stationarity condition with respect to u requires that the Hamiltonian (A4.8) be stationary, and this condition involves no time derivatives for any of the variables. This treatment is, of course, just that of the classic Pontryagin maximum principle.

In the case of linear $a(x,u)$ and quadratic $c(x,u)$ the stationarity conditions will be linear and take the familiar symmetric (self-conjugate) form (15.2).

The path integral (A4.1) will still have the same character for higher-order dynamics and in discrete time; we continue to regard the self-conjugacy of the matrix Φ of the stationarity conditions (c.f. Chapters 10 and 15) as expressing Hamiltonian structure.

To reiterate a point already made repeatedly, the fact that the optimization of a higher-order, stochastic and risk-sensitive problem can be phrased as the free extremization of a path integral, yielding Hamiltonian stationarity conditions, must be regarded as a powerful extension of the maximum principle.

References

Adams, M.B., Willsky, A.G. and Levy, B.C. (1984). Linear estimation of boundary value stochastic processes. *IEEE Trans. Autom. Control*, **29**, 803–821.

Arov, D.Z. and Krein, M.G. (1983). On the evaluation of entropy functionals and their minima in generalised extension problems. *Acta. Sci. Math.*, **45**, 33–50.

Arrow, K.J. (1971). The theory of risk aversion. In *Essays in the Theory of Risk-bearing*, North Holland, Amsterdam.

Bensoussan, A. and Van Schuppen, J. H. (1984). *Proc. 23rd IEEE Conf. on Decision and Control, Las Vegas, NV*, 1473.

Bensoussan, A. and Van Schuppen, J.H. (1985). Optimal control of partially observable stochastic systems with an exponential-of-integral performance index. *SIAM J. Control and Opt.*, **23**, 599–613.

Bertsekas, D.P. (1976). *Dynamic Programming and Stochastic Control*, Academic Press, New York.

Blackwell, D. (1962). Discrete dynamic programming. *Ann. Math. Statist.*, **33**, 719–26.

Bouakiz, M. and Sobel, M.J. (1984). Nonstationary policies are optimal for risk-sensitive Markov decision processes. *Report, College of Management, Georgia Institute of Technology*.

Caravani, P. (1986). On extending linear quadratic control theory to non-symmetric risky objectives. *J. Econ. Dynamics & Control*, **10**, 83–8.

Caravani, P. and Papavassilipoulos, G. (1987). A class of risk sensitive games. *Report R 193, Istituto di Analisi dei Sistemi ed Informatica del CNR*, Rome.

Dym, H. and Gohberg, I. (1986). A maximum entropy principle for contractive interpolants. *J. Functional Analysis*, **65**, 83–125.

Fershtman, C. and Kamien, M.I. (1987). Dynamic duopolistic competition with sticky prices. *Econometrica*, **55**, 1151–64.

Francis, B.A. (1987). *A Course in H_∞ Control Theory*, Springer, New York.

Friedlander, B., Kailath, T. and Ljung, L. (1976). Scattering theory and linear least squares estimation. Part II: Discrete time problems. *J. Franklin Inst.*, **301**, 71–82.

Friedlander, B., Verghese, G. and Kailath, T. (1977). Scattering theory and linear least squares estimation. Part III: The scattering variables and the estimates. *Proc. 1977 IEEE Decision and Control Conference*.

Gihman, T.I. and Skorokhod, A.V. (1971). *The Theory of Stochastic Processes*, Vol. 3, Springer, New York.

Glover, K. and Doyle, J.C. (1988). State-space formulae for all stabilizing controllers that satisfy an H_∞-norm bound and relations to risk sensitivity. *System & Control Letters*, **11**, 167–72.

Graves, R.L. and Telser, L.G. (1967). An infinite-horizon discrete-time quadratic program as applied to a monopoly problem. *Econometrics*, **35**, 234–72.

Grenander, U. and Szegö, G. (1958). *Toeplitz Forms and their Applications*, University of California Press.

Hagander, P. (1973). The use of operator factorisation for linear control and estimation. *Automatica*, **9**, 623–31.
Hannan, E.J. (1970). *Multiple Time Series*, Wiley, New York.
Hewer, G.A. (1971). An iterative technique for the computation of the steady state gains for the discrete optimal regulator. *IEEE Trans. Automat. Control*, **10**, 382–4.
Holt, C., Modigliani, F., Muth, J.F. and Simon, H.A. (1960). *Planning, Production, Inventories and Work-force*, Prentice-Hall, Englewood Cliffs, NJ.
Horwood, J. and Whittle, P. (1986). Optimal control in the neighbourhood of an optimal equilibrium with examples from fisheries models. *IMA J. Math. appl. Med. Biol.*, **3**, 129–42.
Howard, R.A. (1960). *Dynamic Programming and Markov Processes*, MIT Press and Wiley, New York.
Jacobson, D.H. (1973). Optimal stochastic linear systems with exponential performance criteria and their relation to deterministic differential games. *IEEE Trans. Automat. Control*, **AC-18**, 124–31.
Jacobson, D.H. (1977). *Extensions of Linear-quadratic Control, Optimization and Matrix Theory*, Academic Press, New York.
Ježek, J. and Kučera, V. (1985). Efficient algorithm for spectral factorization. *Automatica —J. IFAC*, **21**, 663–9.
Kalman, R.E. (1960). A new approach to linear filtering and prediction problems. *Trans. ASME, Ser. D, J. Basic Eng.*, **82**, 35–45.
Kalman, R.E. (1963). New methods in Wiener filtering theory. In *Proc. Symp. Eng. Appls. Random Function Theory and Probability*, eds F. Kozin and J.L. Bogdanov, Wiley, New York.
Kalman, R.E. and Bucy, R.S. (1961). New results in linear filtering and prediction theory. *Trans. ASME Ser. D., J. Basic Eng.*, **83**, 95–108.
Karp, L. (1984). Higher moments in the linear quadratic gaussian problem. *J. Econ. Dynamics & Control*, **8**, 41–54.
Karp, L. (1986). Methods for selecting the optimal dynamic hedge when production is stochastic. *Amer. J. Agr. Econ.*, **68**, 553–61.
Karp, L. (1987). Risk aversion in a dynamic trading game. *Working Paper No. 404*, Division of Agriculture and Natural Resources, University of California.
Karp, L. (1988). Dynamic hedging with uncertain production. *Int. Econ. Rev.*, **29**, 621–37.
Karp, L. (1989). The endogenous stability of economic systems: the case of many agents. Unpublished MS.
Kuhn, J. (1985). The risk-sensitive homing problem. *J. Appl. Prob.*, **22**, 796–803.
Kumar, P.R. and Van Schuppen, J.H. (1981). On the optimal control of stochastic systems with an exponential-of-integral performance index. *J. Math. Anal. Appl.*, **80**, 312–32.
Lefebvre, M. and Whittle, P. (1988). Survival optimization for a dynamic system. *Ann. Sc. Math. Québec*, **12**, 101–19.
Lehnigk, S.H. (1966). *Stability Theorems for Linear Motions*, Prentice-Hall, Englewood Cliffs, NJ.
Ljung, L., Kailath, T. and Friedlander, B. (1976). Scattering theory and linear least squares estimation. Part I: Continuous-time problems. *Proc. IEEE*, **64**, 131–9.
Morf, M. and Kailath, T. (1975). Square-root algorithms for least-squares estimation. *IEEE Trans. Autom. Control*, **AC-20**, 487–97.
Newton, G.C. (1952). Compensation of feedback control systems subject to saturation. *J. Franklin Inst.*, **254**, 281–86, 391–413.
Newton, G.C., Gould, L.A. and Kaiser, J.F. (1957). *Analytic Design of Linear Feedback Controls*, Wiley, New York.
Plackett, R.L. (1980). Some theorems in least squares. *Biometrika*, **37**, 149–57.

REFERENCES

Pratt, J.W. (1964). Risk aversion in the small and the large. *Econometrica*, **32**, 122–136.
Ross, S.M. (1983). *Introduction to Stochastic Dynamic Programming*, Academic Press, New York.
Silverman, L. (1976). Discrete Riccati equations. *Control and Dynamic Systems*, ed. C.T. Leondes, **12**, 313–86, Academic Press, New York.
Speyer, J.L. (1976). An adaptive terminal guidance scheme based on an exponential cost criterion with application to homing missile guidance. *IEEE Trans. Autom. Control*, **AC-21**, 371–5.
Speyer, J.L., Deyst, J. and Jacobson, D.H. (1974). Optimisation of stochastic linear systems with additive measurement and process noise using exponential performance criteria. *IEEE Trans. Autom. Control*, **AC-19**, 358–66.
Stengel, R.F. (1986). *Stochastic Optimal Control*, Wiley-Interscience, New York.
Telser, L.G. and Graves, R.L. (1968). Continuous and discrete time approaches to a maximization problem. *Rev. Econ. Stud.*, **35**, 307–25.
Telser, L.G. and Graves, R.L. (1972). *Functional Analysis in Mathematical Economics*. University of Chicago Press.
Theil, H. (1957). A note on certainty equivalence in dynamic planning. *Econometrica*, **25**, 346–9.
Van der Ploeg, F. (1983). Risk-sensitive stabilisation policy: complacency and neurotic breakdown. *Cambridge Growth Project Working Paper No. 537*.
Van der Ploeg, F. (1984). Economic policy rules for risk-sensitive decision making. *Zeitschrift für Nationaloekonomie*, **44**, 207–35.
Van Dooren, P. (1981). A generalised eigenvalue approach for solving Riccati equations. *SIAM J. Sci. Stat. Computing*, **2**, 121–135.
Vostrý, Z. (1975). New algorithm for polynomial spectral factorization with quadratic convergence I. *Kybernetika*, **11**, 415–22.
Vostrý, Z. (1976). New algorithm for polynomial spectral factorization with quadratic convergence II. *Kybernetika*, **12**, 248–59.
Whittle, P. (1951). *Hypothesis Testing in Time Series Analysis*, Almquist and Wicksell, Uppsala.
Whittle, P. (1953). The analysis of multiple stationary time series. *J. Roy. Statist. Soc., B*, **15**, 125–39.
Whittle, P. (1963). *Prediction and Regulation*, English Universities Press, London. (See also Whittle, 1983b.)
Whittle, P. (1981). Risk-sensitive linear/quadratic/Gaussian control. *Adv. Appl. Prob.*, **13**, 764–77.
Whittle, P. (1982). *Optimization over Time*, Vol. 1, Wiley, Chichester.
Whittle, P. (1983a). *Optimization over Time*, Vol. 2, Wiley, Chichester.
Whittle, P. (1983b). *Prediction and Regulation*, second and revised edition of Whittle (1963), University of Minnesota Press and Blackwell, Oxford.
Whittle, P. (1986a). The risk-sensitive certainty-equivalence principle. In *Essays in Time Series and Allied Processes*, ed. J. Gani, 383–8, Applied Probability Trust, Sheffield.
Whittle, P. (1986b). *Systems in Stochastic Equilibrium*, Wiley, Chichester.
Whittle, P. (1989a). A class of decision processes showing policy-improvement/Newton-Raphson equivalence. *Probability in the Engineering and Informational Sciences*, **3**, 397–403.
Whittle, P. (1989b). Entropy-minimising and risk-sensitive control rules. *System and Control Letters*, **13**, 1–7.
Whittle, P. (1990). The stochastic maximum principle. Submitted to *System and Control Letters*.

Whittle, P. and Gait, P. (1970). Reduction of a class of stochastic control problems. *J. Inst. Math. Appl.*, **6**, 131–40.

Whittle, P. and Komarova, N. (1988). Policy improvement and the Newton–Raphson algorithm. *Probability in the Engineering and Informational Sciences*, **2**, 249–55.

Whittle, P. and Kuhn, J. (1986). A Hamiltonian formulation of risk-sensitive linear/quadratic/gaussian control. *Int. J. Control*, **43**, 1–12.

Wilson, G.T. (1969). Factorization of the covariance generating function of a pure moving average process. *SIAM J. Numer. Anal.*, **6**, 1–7.

Wilson, G.T. (1972). The factorization of matrical spectral densities. *SIAM J. Appl. Math.*, **23**, 420–26.

Zames, G. (1981). Feedback and optimal sensitivity: model reference transformations, multiplicative seminorms and approximate inverses. *IEEE Trans. Autom. Control*, **AC-16**, 585–601.

Index

Abbreviations, 232

Backward translation operator, 7
Blackwell policy, 220

Canonical factorization, *see*
    Factorization, canonical
Certainty-equivalence principle, 25 *et seq.*, 37, 60
  risk-sensitive, 81 *et seq.*, 95, 135
Closed-loop control, 10, 20
Companion matrix, 9
Conjugate variables, 69 *et seq.*, 131, 134
Controllability, 39, 54
Control-power matrix, 40, 98
Cost function, 2

Derivate characteristic function, 229
Detectability, 64
Deviation sensitivity, 39
Discounting, 221
Discrepancy, 81
Dual variables, *see* Conjugate variables
Duality, of past and future, 67 *et seq.*
Dynamic programming equation, *see*
    Optimality equation

Entropy criterion, 91, 209 *et seq.*
Estimate
  least square, 22
  maximum likelihood, 22
  minimum stress, 83
Estimation, 20 *et seq.*, 237 *et seq.*
Euphoria, 88, 105
Excitability, 63
Exogeneity, 17

Factorizability, 154
Factorization
  canonical, 10, 44, 73, 141, 149, 52 *et seq.*, 173 *et seq.*, 199 *et seq.*
  iterative methods for, 166 *et seq.*
  reduced, 117
Factorization/Riccati relation, 165, 193 *et seq.*
Feedback/feedforward, 36, 46, 101
Free form, 84

Gain matrix, 34, 9, 110, 148
  predictive, 99, 110
Gauss–Markov Theorem, 10, 22, 237

Hamiltonian formulation or formalism, 43 *et seq.*, 72 *et seq.*, 131 *et seq.*, 239
Higher-order models, 6, 161 *et seq.*, 200 *et seq.*
History, 15
Horizon, 15, 20
Horizon-stability, 34, 39, 54 *et seq.*, 118 *et seq.*, 154 *et seq.*, 179 *et seq.*
H_∞ criterion, 11, 90, 208

Information, *see* Observables
Innovation, 23, 61

Kalman filter, 61 *et seq.*, 65, 66, 73, 75, 108, 109

Large-deviation theory, 229
Least square (or linear least square) estimate, 22
Legendre transform, 41
LEQG structure, 4, 15 *et seq.*, 79 *et seq.*
LQ-approximation, 56
LQG structure, 3, 15 *et seq.*

Maximum likelihood estimate, 22
Maximum principle, 10, 41 *et seq.*, 132
  stochastic, 10, 133, 140, 229
Minimum stress estimate, 83
Multi-agent models, 225

INDEX

Neurosis, 87
Newton–Raphson method, 49 *et seq.*, 124, 167, 169, 197
Noise vector, 16
Non-LQG models, 222, 229
Notation, 234

Observability, 64
Observables, 16, 18
Observation noise, 2
Optimality equation, 18 *et seq.*, 31, 229
Optimism, 87 *et seq.*

Path-integral formalism, general, 143 *et seq.*, 147 *et seq.*, 191 *et seq.*
Path-integral formulation, 131 *et seq.*, 147, 189
Pessimism, 87 *et seq.*
Plant equation, 1
Policy, 2
Policy improvement, 49 *et seq.*, 122 *et seq.*, 155 *et seq.*, 166 *et seq.*, 169, 180 *et seq.*, 197 *et seq.*
Pontryagin maximum principle, *see* Maximum principle
Prediction, 7, 74, 85, 138
Process noise, 2
Process variable, 1, 3
Projection, 23

Realizability, 19
Recoupling, 110 *et seq.*, 185 *et seq.*
Reference signal, 18
Regulation, 32
Riccati equation, 33, 40, 53, 62, 66, 98, 102, 108, 109, 148
  numerical solution of, 48 *et seq.*

Riccati/factorization relation, 165, 193 *et seq.*
Risk averseness, 5, 87
Risk-resistance condition, 99
Risk seeking, 5, 87
Risk-sensitive certainty-equivalence principle, 81 *et seq.*, 95, 135
Risk sensitivity, 4, 87
Risk-sensitivity parameter, 5
  break-down values of, 88, 111, 114 *et seq.*, 225

Saddlepoint, 81, 82, 88
Separation principle, 28, 84, 95
Set points, 37
Stabilizability, 39
Stability matrix, 34
State structure, 9, 20, 31
State variable, 1, 20
Stochastic maximum principle, 10, 133, 140, 229
Stress, 81, 132, 134
  future, 94, 96 *et seq.*
  past, 94, 107 *et seq.*
Survival, 224
System formulation, 138, 144

Time homogeneity, 1, 4
Time invariance, 4
Transfer function, 11
Translation operator, 7

Utility, 5

Value function, 19, 31, 162 *et seq.*
Value iteration, 49, 114

White noise, 3
Wiener filter, 7, 10, 74

Books are to be returned on or before
the last date below.